花都

吃 玩 住 赏 游 购

WONDERFUL TRAVELLING EXPERIENCE IN NEW HUADU

广州市花都区旅游局 编

在版图书（CIP）数据

花都攻略 / 广州市花都区旅游局编. —— 广州：广东旅游出版社，2013.9
ISBN 978-7-80766-568-7

Ⅰ. ①花… Ⅱ. ①广… Ⅲ. ①区（城市）—旅游指南—广州市 Ⅳ. ①K928.965.1

中国版本图书馆CIP数据核字（2013）第166882号

责任编辑：何　阳　梅哲坤
英文编辑：梁嘉韵
美术编辑：邓传志　谢晓丹
责任校对：李瑞苑
责任技编：刘振华

本书图片由广州市花都区旅游局、广州市花都区宣传部共同提供。
图片摄影：李八一　关振伦　和匀生　姚日文

广东旅游出版社出版发行
（广州市越秀区先烈中路76号中侨大厦22楼D、E单元）
邮编：510095
邮购电话：020-87348243
广东旅游出版社图书网：www.tourpress.cn
广州汉鼎印务有限公司
（广州市天河区棠下高沙工业区广棠路21-23号）

889毫米×1194毫米　32开　5印张　85千字
2013年第1版第2次印刷
印数：15000册
定价：35.00元

【版权所有　侵权必究】

本书如有错页倒装等质量问题，请直接与印刷厂联系换书。

WELCOME TO HUADU

请到花都来·行走花都
畅游天地间之风雅灵秀

花都旅游资源在方圆远近之竞技场中

很可能就是全能冠军

山水湖泊、人文历史，一个都不会少

Huadu tourism resources in the competition of the near or distant place
May be the all-around winner
Lakes, landscape and art history, all are included

杨伊琳，出生于广州市花都区狮岭镇杨二村，中国女子体操队员。2003年3月加入广东省队，担任广东队队长。2007年1月10日进入国家队。2008年体操世界杯天津站高低杠冠军，自由操亚军，2008年法国邀请赛全能、高低杠冠军，2008年北京奥运会为中国代表团夺得了1枚金牌、2枚铜牌（高低杠、个人全能）。2010年广州亚运会暨第16届亚运会团体冠军，2010年11月9日，杨伊琳担任广州亚运会火炬传递广东省广州市首棒火炬手。

Yang Yilin was born in the Yanger village, Shiling Town, Huadu District, Guangzhou City, the Chinese women's gymnastics team member. In March 2003, she joined the Guangdong team, and served as the captain of the Guangdong team. On January 10, 2007 she entered the national team. She won the uneven bars champion and the floor exercise runner-up at the 2008 Gymnastics World Cup Tianjin Stop, all-around and uneven bars champion at the 2008 French Invitational, and won gold and two bronze medals (uneven bars, individual all-around) for the Chinese delegation at the 2008 Beijing Olympic Games. She was the team champion at the 2010 Guangzhou Asian Games & the 16th Asian Games. On November 9, 2010, Yang Yilin was the first torchbearer of Guangzhou Asian Games torch relay for Guangzhou City, Guangdong Province.

个人简介
PERSONAL PROFILE

杨 伊琳
YILIN YANG

花都旅游形象大使
IMAGE AMBASSADOR OF HUADU TOURISM

新花都·新旅游·天地有大气

花都区位于广东省中南部,珠江三角洲的北端,扼守广州的北大门,是南北交通要道,素称"省城之屏障,南北粤咽喉"。随着交通大格局的规划及建设,汽车与空港优势让花都不仅以优异的山水资源成为"广州后花园",更兼备了广州"第一会客厅"的实质城市功能。

远古时期,盘古王开天地,在一片混沌之中,盘古王凭借其一夫之力,为人类支撑出了一片全新的天地,如今我们走动的世界,也是他的伟力所成。花都自古以来,便有崇拜盘古王的习俗,他成为花都人的偶像,也是当代花都人的真实写照。

花都人承继了盘古王开天辟地、敢为人先、敢担天下道义的精神,这种精神是历史积淀,也是现代品质,更是当前广东省提倡的"厚于德、诚于信、敏于行"的真正体现。

花都始祖,在这片"三山一水六平原"土地上开疆辟土,百折不挠,世代勤劳智慧、与山水和谐共有,当代人民顺势而为、奋发图强,在古"花县"这块神奇美丽的土地上,成就了中华文明的重要一隅。不管是洪秀全的革新,还是现如今花都人的更高、更快、更强的产业链,都是花都人志存高远、敢拼敢闯的智慧表现,这也是花都人谋局而动、创新不止的崇高精神。

为更好地宣传和推介花都,让更多人分享花都的山水与精神真髓,花都区政府、区旅游局努力,在极短的时间内,编纂了《花都攻略》,希望各方游客,一踏足岭南之地之首站——花都,便可从不同角度走近花都、了解花都。

——中共广州市花都区委副书记、广州市花都区人民政府区长

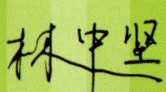

WONDERFUL TRAVELLING EXPERIENCE
IN NEW HUADU

Located in the south-central part of Guangdong Province, northern end of the Pearl River Delta and in the front of northern part of Guangzhou, Huadu is the geological artery of the north-south traffic and is known as "Shield of Provincial City, Fortress of South-North-Guangdong Province". With the planning and construction of the traffic pattern, land-air-port advantage makes Huadu to be the "Back Garden of Gangzhou" because of excellent landscape resources, which also possess the virtual city function of "First Reception Hall" of Guangzhou

Before the ancient times, King of Pan Gu created the heaven and the earth. In the chaos, King of Pan Gu, by himself, supported a brand new world for human beings. The world we live in is created by him. Since the ancient time, Huadu has had the tradition of worshiping the King of Pan Gu, who is the idol of local inhabitants in Huadu, being the true portrayal of the contemporary local inhabitants.

People in Huadu have inherited the pioneering spirit and the courage to assume the world moral from King of Pan Gu, which is the historical accumulation and modern quality, even the true demonstration of "Kindness on the moral, credibility on the credit and quickness on the action" advocated by the Guangdong Province.

The ancestors of Huadu cultivated virgin land with their perseverance, diligence and wisdom of generations, creating a harmonious landscape. The contemporary people work hard on this ancient, mysterious and beautiful "Huaxian County" based on their ancestor's rich fruits, playing an important role in the Chinese civilization. Whether the innovation of Hong Xiuquan, or the higher, faster and stronger industrial chain of modern people, is the wisdom reflection of great ambitions and courage to create new change of Huadu people, which is also the lofty spirit of action through plan and long-lasting creativity.

In order to better promote and introduce Huadu to more people to enjoy her landscapes and spirit, the government of Huadu District and Tourism Administration of Huadu District strive to compile the travel guide of Huadu in a short period. Hope the visitors and guests home and abroad, from the first station of Lingnan----Huadu, to approach Huadu and learn Huadu from different perspectives.

—— Deputy Secretary of Huadu District Committee of Guangzhou of China, District Mayor of People's Government of Huadu District of Guangzhou

Lin Zhongjian

CONTENTS 目录

邀请函 / 10
Invitation Letter

印象花都 / 12
Impression of Huadu

花都历史沿革 Historical evolution of Huadu
花都生态特写 Ecological feature of Huadu
花都人文特写 Cultural feature of Huadu
花都物候特写 Living beings feature of Huadu
花都节庆、习俗 Festivals and conventions in Huadu
花都物候特产 Specialty in Huadu

花都全攻略 / 31
Overall Guides for Traveling in Huadu

新华街 Xinhua Street
花山镇 Huashan Town
花东镇 Huadong Town
赤坭镇 Chini Town
炭步镇 Tanbu Town
狮岭镇 Shiling Town
梯面镇 Timian Town
雅瑶镇 Yayao Town

特色主题线路推荐　　/ 137
Distinctive Itinerary Recommendation

畅游花都，你还可以这样玩！
Entertainment recommended for travel in Huadu

花都酒店介绍　　/ 141
Hotel Introduction in Huadu

交通全攻略　/ 153
Communication of Huadu

后记　/ 156
Postcript

附：花都区旅行社（门市部）名录
Attached: Directory of Travel Agencies
(Sales Department) in Huadu District

INVITATION LETTER 邀请函

花都欢迎您

"不要为生命忧虑吃什么、喝什么,为身体忧虑穿什么,你们看那天上的飞鸟,也不种,也不收,也不积蓄在仓里,你们的天父尚且养活他,你们不比飞鸟贵重得多么?"

当行色匆匆的生活充彻着我们生命里的每一天的时候,当我们为了房子、为了孩子、为了钞票、为了职位、为了"强取豪夺"、为了"蝇头小利"而营营碌碌地生活,难道这就是一辈子的终极目标?

我们全然忘记了,来一趟人世间多么的偶然和不容易,为什么要让这些坏习惯、坏毛病填充我们的生活?我们身边如斯美妙,值得我们停下脚步来欣赏的东西太多太多,你看那春有油菜花;夏有荷叶田田、莲花井蒂;秋有硕果累累、稻穗满贯;冬有桑果满园、芋头飘香;还有那先人的智慧、古民居的静谧,花都之美不仅在于山野之灵秀,还在于现代业的纷繁。某些乡镇,美的气息如滔滔江水绵绵不绝;而某些地方,花都之美是在坚固的古建筑群落之中,在那些貌似颓垣的瓦砾却丰盈无比的寓意之中。那些美,更能勾起惊心动魄之探窥之欲望。

就找一天,花点时间,从容地停留,你会发现,山水灵秀、人杰地灵的花都,将是你身心灵的修持之地,花都在你不远的身边和心间。你将发现,你生命中所真正需要的,比如休憩、比如美食、比如路向、比如信仰,甚至,比如呼吸吐纳之大气,都能在花都之地找寻到适合你的方式。

那一天,就为那一天,请到花都来。

——广州市花都区旅游局

WELCOME TO HUADU

Therefore I tell you, do not worry about your life, what you will eat or drink; or about your body, what you will wear. Is not life more important than food, and the body more important than clothes? Look at the birds of the air; they do not sow or reap or store away in barns, and yet your heavenly Father feeds them. Are you not much more valuable than them?

Have you ever stopped your hurried footsteps to review your life? What goals are you striving for in your life? House? Children? Money? Or post? Are you busy working all the day just for achieving these goals?

Actually we totally ignored the real meaning of our lives. What a coincidence and luck for us to live in this beautiful world! There are so many beautiful scenery and wonderful things for us to enjoy. Why should we waste our precious time and vigor to pursue those material benefits? Come to the wild of Huadu and enjoy the colorful natural scenery in the four seasons. You may enjoy the rape flower in the spring, lotus and water lily in the summer, fruits and rice in the autumn, and mulberry fruits and taro in the winter. You'll be impressed by the wisdom of the ancient people and the tranquil ancient folk houses. The beauty of Huadu lies not only in its landscape but also in its flourishing modern industry. In some towns, you can find some ancient buildings, which have special beauty and can easily arouse people's curiosity and interest.

Why don't you find some time and have a trip to Huadu? You'll find it's such a wonderful place for cultivating your mind and character. You can feel the beautiful mountains and rivers coming into your heart and body. In this place, you can attempt to do everything that you wish to do in your life, such as resting in the woods, tasting delicious foods, finding your belief and living direction, and breathing fresh air.

Friends, welcome to Huadu and enjoy wonderful days for yourself!

—— Guangzhou Huadu District Tourism Administration

IMPRESSION OF HUADU 印象花都

花都历史沿革
Historical evolution of Huadu

现用名：花都区

原名：花县

汉朝：属番禺管辖

隋朝：属南海县辖

宋朝后：分属番禺、南海县辖

清康熙二十五年（公元1686年）：取南海、番禺两县部分区域置县，因县城近花山，定名"花县"，属广州府

民国时期：花县主要由省直辖

建国后：隶属江北专区、珠江专区、粤北行政区、佛山专区

1960年4月：划为广州市属县

1993年6月18日：经国家批准，撤县设市（县级），撤销花县，定名为"花都"，由省人民政府委托广州市代管

2000年5月21日：经国家批准撤市设区，成为广州市花都区

Present name: Huadu District   Previous name: Huaxian County
Han Dynasty: under the jurisdiction of Panyu
Sui Dynasty: under the jurisdiction of Nanhai County
After Song Dynasty: under the joint jurisdiction of Panyu and Nanhai County
In 25th year of Emperor Kangxi in Qing Dynasty (A.D. 1686): selected the parts of Nanhai County and Panyu County as the "county", the county town was named "Huaxian County" because it was near the Huashan Mountain and was under the jurisdiction of Guangzhou Prefecture (special term in Qing Dynasty)
During the Republic of China: the Huaxian County was under the direct jurisdiction of the province
After the foundation of People's Republic of China: under the jurisdiction of Jiangbei Prefecture, Zhujiang Prefecture, Administrative Region of North Guangdong and Foshan Prefecture
In April, 1960: it was designated as the municipal country of Guangzhou
June 18th, 1993: approved by the government, withdraw the "county" and set up "city" (equal to county), "Huaxian County" was withdrawn, renamed as "Huadu" and under the jurisdiction of Guangzhou, entrusted by the provincial people's government
May 21st, 2000: approved by the government, "city" was replaced by "district", being the Huadu District of Guangzhou

花都生态特写
ECOLOGICAL FEATURE OF HUADU

生态的花都，值得大书特书。

俗话说，花都里有"三山一水六平原"。花都虽然没有名山大川，但不乏秀丽妩媚的山水景观。有近200平方千米的山地，湖泊、水库众多，建有中小型水库17座，知名的水库有芙蓉嶂水库、九湾潭水库、三坑水库、福源水库、集益水库等。环绕水库的是森林覆盖率高、林相整齐的秀丽山峰。森林公园也是花都一绝，九龙湖森林公园、王子山森林公园、丫髻岭森林公园等闻名遐迩。

长期生活在钢筋丛林中，呼吸着大量汽车尾气、工业废气、空调排气；长时间呆在不通风的空调房间内，人们经常会觉得口干舌燥、咽喉肿痛、头晕犯困。城市中巨大的生活压力，让人的精神也无时无刻不处在一种紧绷的状态。有什么办法能让我们放松下来，吸氧洗肺、放松心情、健体强身，让下一周的生活更有活力？那便是"森林浴"。去森林中呼吸芳香清新的空气，让大自然帮助自己来一次呼吸道的清洁润滑。

目前，花都区林业用地面积57.5万亩，占国土面积的39.7%，拥有这样的生态内涵，花都人怎能不幸福？

Ecological Huadu is worthy of detailed introduction.

As the saying goes, there are "Three mountains, one lake and six plains" in Huadu. Although there are no famous mountain and gorgeous river in Huadu, there is no lack of beautiful, charming and natural landscape. There are nearly 200 square kilometers of mountain land and plenty of lakes and reservoirs. 17 small and middle-sized reservoirs are built there, of which the famous reservoirs include: Fulongzhang Reservoir, Jiuwantan Reservoir, Sankeng Reservoir, Fuyuan Reservoir and Jiyi Reservoir etc. The reservoir is surrounded by pretty peaks with high forest coverage rate and regular form. The forest park is one special feature of Huadu and Dragon Lake Forest Park, Prince Mountain Forest Park as well as Yajilin Forest Park are renowned in the world.

Live in the steel jungle for long time and breathe in a lot of vehicle exhaust, industrial emission and air-conditioning exhaust; long stay in the air-conditioned room without any ventilation will cause thirst, sore throat and dizziness and drowsiness. Human beings are always in a state of tension because of tremendous life pressure in the city. It there any way to make us relax, breathe the oxygen and clear the lung, strengthen our physique and become more dynamic in the next week? "Forest Bathing" is a good choice. Go to the forest to breathe fresh air and the nature will help us clean and lubricate our respiratory tract.

Currently, Forestry land of Huadu District covers an area of 575,000 mu, accounting for 39.7% of the national territorial area. We will be happy when we live in the ecological Huadu.

IMPRESSION OF HUADU 印象花都

花都人文特写
Cultural feature of Huadu

　　花都，人文荟萃。从明清时代留存下来的诸多古村落、古建筑群、古书院等，便可略见一斑。尽管目前仍未有相关方面专家的考察结论，但花都人芳华自知。

　　花都区的历史文化旅游资源丰富多彩，有洪秀全故居、洪秀全纪念馆、资政大夫祠古建筑群、盘古王庙、炭步镇塱头村民居古建筑群、花东镇高溪村欧阳庄民居古建筑群等，另外还有宗教文化建筑的圆玄道观，其建筑雄伟，是省内首屈一指的道教圣地。在历史文化旅游资源中，花都作为太平天国领袖洪秀全的故乡，在风卷云舒的历史画卷上留下了浓墨重彩的一笔，也为花都旅游的今天写下重要的一笔。

Huadu, is a gathering place of advanced culture and talents. We can catch a glimpse from the ancient villages, ancient architectural complex and the ancient academy survived from the Ming Dynasty and Qing Dynasty. Though currently, there are no relevant conclusions, people in Huadu, they know the beauty and charm of Huadu.

Huadu District is abundant in historical and cultural tourism resources such as Former Residence of Hong Xiuquan, Memorial Hall of Hong Xiuquan, Ancient Architectural Complex of Ancestral Temple of Zizheng Senior Official (an official position in Yuan, Ming and Qing Dynasty), Temple of King of Pan Gu, Ancient Architectural Complex of Folk House of Langtou Village of Tanbu Town, Ancient Architectural Complex of Folk House of Ouyangzhuang of Gaoxi Village of Huadong Town etc. In addition, there is also Yuen Yuen Taoist Temple with religious culture, which is the premier Taoist shrine of the province with gorgeous architecture. In the historical and cultural tourism resources, Huadu, as the hometown of Hong Xiuquan, the leader of the Taiping Heavenly Kingdom, leaves such an indelible mark on the eventful history and on the tourism of Huadu.

花都物候特写
Living beings feature of Huadu

　　俗话说，一方水土养一方风物。花都肥沃的土壤上，不仅人杰地灵，更是地大物博。

　　在花都，物候、物种奇特而别致，是休闲、度假旅游的又一亮点。

　　颇有观赏价值的花山菊花石、赤坭瑞岭盆景及炭步香芋，北兴荔枝、李溪龙眼、锦山红柿、京塘莲藕等名特产品，自古以来，或是朝廷贡品，或是特优产品，都有响当当的名号，但可惜，多年来，这些物候老字号犹如藏于深闺，仍未被发掘，成为民间瑰宝。

　　花都在美食方面，可圈可点的地方太多太多，乃至在休闲度假时光中，来自"美味之乡"广州的人来到花都，流行打包带回广州，可见，"美食在广州，味道在花都"。

　　当前，尤为值得称赞的是花都的"十佳农家菜馆"，在青山绿水间，在城郊乡下，总有别具特色的农家乐点缀其中。吃腻了高档酒楼里的精雕细琢，再到农家乐里吃地道的农家小菜，粗茶淡饭，土鸡土鸭，甚至山珍海味，或是带着浓郁的"童年的味道"、"奶奶的味道"的家乡菜，那滋味实在是再多金子都换不来的。

　　花都物候如此丰盛，这日子怎能不幸福？

　As the saying goes, each place has its own way of supporting its own living beings. Huadu not only has outstanding people, but also has vast and rich resources.

　In Huadu, the living beings and species are unique and original, being the highlight of relaxation and vacation tour.

There are chrysanthemum stone of Huashan Mountain with high ornamental value, Miniascape of Ruiling Village of Chini Town, Tanbu Taro, Beixing leechee, Lixi Longan, Red Persimmon of Jinshan Village and Jingtang Lotus etc in Huadu, since ancient times, some of which are imperial tribute, some are awarded the products of excellence, all of which have great charm. Unfortunately, over the years, these traditional and reputable articles, as if hidden in the purdah, have not been discovered, becoming folk treasures.

On terms of gourmet, there are so plenty of famous dishes that during the holidays, many people from Guangzhou—"Home of Gourmet" come to Huadu and bring the delicious food back. We can see that "Gourmet is in Guangzhou, Flavor is in Huadu".

Currently, the "Top 10 Farm Food Restaurants" in Huadu are praiseworthy. There are always the distinctive happy farmhouses either among the green hill and pure water or in the suburb or countryside. Tired of rich food of the high-end restaurant, you enjoy the various kinds of farm food, though not as delicate as the rich food, it can recall your memory. Tasting dishes with the "Childhood memory" or "Grandma's flavor" is an unforgettable experience.

With such an rich resources, we must be happy!

IMPRESSION OF HUADU 印象花都

① 看花都人一年怎么过
How Huadu people spend one year

春节，斗鸟过年（Spring Festival —— bird fighting during the festival）

春节，"过年"，有扫尘、贴春联、挂年画、放鞭炮、拜年、派利是、舞狮、逛花街等习俗，从农历十二月二十四日谢灶爷后便入年关，至正月十五日元宵节才出年关。至今多数保留煮煎堆、炸油角、炒米饼等习惯。除夕晚饭前，各家各户要贴贺春对联、门神年画，祭祀祖先，迎接灶君回来。全家人都要赶回来吃晚饭，俗称"哞年饭"，以庆祝全家平安度过一年。除夕过了12点，即新的一年开始。守岁的人欢呼雀跃，燃放鞭炮、烟花，向后辈"派利是"（发红包）。按照预先择定的吉时，联宗到祠堂举行迎春祭祖，家家户户祭祀神灵。初二这天，家家户户准备猪肉、鸡鹅、酒水，择吉时联宗到祠堂祭祀祖先；各家各户在家门口祭祀神灵。全家聚餐，席上必有鸡、蚝豉（好市）、发菜（发财）等。出嫁女要回娘家"担节"，与父母兄弟团聚。

如果你觉得上述的民俗仍不够"花都化"的话，那"斗鸟过年"应该算是最有花都特色。春节期间，花都区的花山镇民间会举办"新春斗鸟大赛"，几十只画眉鸟上演一幕幕精彩的"笼中斗"。"斗画眉"这项民间活动源于宋朝，花山每年过年都会举办斗鸟大赛，这已是一项传统。

Spring Festival, "Celebration of Spring Festival", there are many conventions such as dust sweeping, spring festival couplets posting, spring festival pictures posting, setting off the firecrackers, give New Year's greetings, giving red packets, lion dancing and wander in the traditional street etc. From the 24th of twelfth lunar month, the beginning date into the end of the year, on which day people express thanks to the Kitchen God, until the finish on the Lantern Festival on 15th of first lunar month. So far majority of conventions are retained such as cooking the fried glutinous rice balls with sesame, frying the bread dumplings and frying the rice cake etc. Before the New Year' Eve dinner, each family will post the spring festive couplets and spring festival pictures of Door God, offer sacrifice to the ancestors and welcome to the Kitchen God. All family members should come back to have the dinner, which is called "Family Reunion Dinner", in order to celebrate all family member's safety in the previous year. After twelve o'clock of the New Year's Eve, a new year begins. Staying-up people cheer up, set off the firecrackers and give the red packets to the younger generations. The clansmen go to the ancestral hall to welcome the spring and offer sacrifice to the ancestors on the auspicious time. Each family will offer sacrifice to the ancestors. On the second day of first lunar month, each family prepares pork, chickens and geese and drinks, the clansmen go to the ancestral hall to offer sacrifice to the ancestors on the auspicious time selected in advance; they also worship the god at home. Chicken, dried oyster (booming market) and hair weeds (being rich) are must in the family reunion dinner. The married women will go back home, give away gifts and get together with the family.

花都节庆、习俗
FESTIVALS AND CONVENTIONS IN HUADU

If you think that the conventions above should not be called distinguishing feature of Huadu, "Bird Fighting during the Spring Festival" is the unique feature of Huadu. During the Spring Festival, the folk people of Huashan Town, Huadu District will organize the "Bird Fighting Competition of New Spring", dozens of wood thrushes fight with each other in the cage. "Wood Thrush Fighting", the folk activity, originated from Song Dynasty and is hold by Huashan Town every year during the Spring Festival, being a tradition.

闹元宵，舞犁耙（Celebration of Lantern Festival, Dancing with Plow Harrow）

元宵节，俗称上元节或灯节，花都有"元宵罢，舞犁耙"的说法，家家户户吃汤圆或当地特有的"圆仔"，上一年生了男孩的家庭都要"起灯（丁）"、"摆灯（丁）"酒，而亲友赠送小孩布料、衣物、猪肉等"贺灯（丁）"。晚上，还要挂灯笼，并提灯笼（如鲤鱼灯笼、蚌虾灯笼等）游行。

Lantern Festival is also called Shangyuan Festival or Festival of Lanterns. There is a kind of convention "Celebration of Lantern Festival, Dancing with Plow Harrow" in Huadu. Every family eats the rice dumpling or local "Round Dumpling" of Huadu. The family with boy born in the last year will "post the lantern (boy)", "prepare the lantern feast" while the relatives and friends give away the baby cloths and pork to "celebrate the lantern (boy)". In the evening, people hang the lanterns (Carp-shape lantern, Clam-shape lantern and Shrimp-shape lantern) and participate in the lantern parade.

清明节（Tomb-sweeping day）

在每年的农历三月初三日前，即公历4月5日左右，中国人有祭祖扫墓、缅怀先人的习俗。花都人的旧俗里，家家户户在门口插招魂柳或插艾草、丝毛草、桃叶等，以示辟邪。是日，花都家家户户吃粉角，俗称"清明角"，"闭墓"（由清明至闭墓为期一个月）当天，花都人有吃"裹叶糍"的习惯。在这为期一个月的时间中，人们必须到祖先墓地祭祀，清扫墓园，压"山钱"，祭品主要为"清明角"、"清明蔗"、煎咸鱼、咸鸭蛋和烧肉等。

Whenever on the third day of third lunar month each year, which is around the 5th of April of the Gregorian calendar, people will worship the ancestors, sweep the tomb and recall the ancestors. In the old convention of Hudu, people will insert the willow of eaves, wormwood, couchgrass and peach leaf to exorcise evil spirits. Each family in Huadu eats the fried rice dumplings, which are called "Qingming Dumpling". On the Ending Day of Tomb Sweeping (one month from the Tomb-sweeping Day to the Ending Day of Tomb Sweeping), people will eat the "Wrapped Rice". During that one month, people must go to the tomb of ancestors and offer sacrifice to the ancestors, sweep the tomb, give the ghosts money. The offerings include "Qingming Dumplings", "Qingming Sugarcrane", fried salted fish, salted duck egg and carbonado etc.

IMPRESSION OF HUADU 印象花都

① 看花都人一年怎么过
How Huadu people spend one year

端午节（Dragon Boat Festival）

每年农历五月初五，花都人有包粽子的习惯。而且外嫁女必须在节前拿着粽子和粉面等礼物回娘家，与父母兄弟联络感情，此俗为"担粽"。节日当天，人们以粽子酬谢神灵并许愿。

On the fifth day of the fifth lunar month each year, people of Huadu make rice dumplings. The married woman shall take the rice dumplings and noodles and other gifts back to parents' home and get together with other family members, this convention is called "sharing rice dumplings". On that day, the rice dumplings are also used to express thanks to the God and make a vow.

盂兰节（Yulan Festival）

又称"中元节"、"阴节"、"鬼节"，是祭祀孤魂野鬼的节日。传说七月十四日这天，阎罗王放出无主孤魂到处寻找衣食，故村民不让小孩当日晚上到野外走动，害怕他们遇上孤魂野鬼。

It is also called "Hungry Ghost Festival", "Infernal Festival" and "Ghost Festival", which is a festival to offer sacrifices to ghosts. It is said that on the fourteenth day of seventh lunar month, the Yama releases ownerless souls to look for clothing and food everywhere. So villagers don't allow their children to walk around in the evening in case of coming across of ghosts.

中秋节（Mid-Autumn Festival）

又叫"月光诞"，每年的农历八月十五日晚，村民准备鸡、猪肉和酒水，全家人团聚用餐。饭后有赏月活动，旧时多在户外挂"走马灯"或竹扎彩灯，摆设香案拜（赏）月，案上有月饼、水果、田螺、芋头和三牲肉类等，烧香点烛遥拜月亮（称之"拜月光"）。节前，外嫁女要送月饼、猪肉等礼物回娘家，称之"担月饼"。

花都节庆、习俗
FESTIVALS AND CONVENTIONS IN HUADU

It is also called "Birthday of Moon". On the fifteenth evening of eighth lunar month each year, villagers prepare chicken, pork and drinks for the family reunion dinner. People will enjoy the moon after the dinner. In the old times, people will hang the "Trotting Horse Lamp" or colorful lamp made of bamboo in the open air and furnish incense burner table to worship the moon, on which there are moon cakes, fruits, escargots, taros and meat of three kinds of animal etc. Before burning the incense and worshiping the moon (which is called "Moon Worshiping"), the married woman should take moon cakes and pork back to parents' home, which is called "sharing moon cake".

重阳节（Chung Yeung Festival）

又叫重九节、菊花节，每年的九月初九日，村民会在这天登高转运，寓意步步高升，祈祷转个好运。比起广东省的其他区域，花都人对重阳节尤为重视，青年在登高处放风筝取乐，小孩则买风车"转运"。为防山火，现在政府指定此节为"敬老节"，多数村委会将全村的老人请在一起聚餐，派礼品。这一天，也是民间祭扫祖坟的日子。

It is also called Double Ninth Festival and Chrysanthemum Festival. On the ninth day of ninth lunar month each year, villagers will ascend a height to change the fortune, which means to attain eminence step by step and looking forward to good fortune. Compared with other area of Guangdong, people of Huadu pay more attention to this festival. The youth on the high place fly kites and children buy the small windwill to "change the fortune". In order to prevent the fires, the government only opens the Mount of King of Pan Gu, Furong Peak and Timian gaobaizhang to the villagers, other mountains are prohibited from climbing .

This festival has evolved into "Festival for the Aged". Many village committees get the aged together for dinner and giving away gifts. This day is also a day for worshiping the ancestor's tomb.

冬节（Winter Soltice）

冬至，又叫"过冬"、"冬节"。每年的农历十一月下旬冬至日，村民备好三牲肉类、香烛等，拜祭天下众神，花都人习惯在这一天煮汤圆或"圆仔"，取团圆、甜蜜之意，晚餐如同过年一样丰盛，俗称吃"团冬饭"。

The Winter Solstice, is also called "Hibernating" and "Potlatch". On the Winter Solstice, end of the eleventh lunar month each year, villagers prepare the meat of three kinds of animals and joss sticks and candles to worship the god of the world. People of Huadu have got accustomed to eat the rice dumpling or "Round Dumpling" to look forward to the reunion and sweetness. The dinner is as rich as the dinner of New Year's Eve, which is called "Reunion Dinner of Winter Solstice".

IMPRESSION OF HUADU 印象花都

❷ 看花都人的特别习俗
Special custom in Huadu

花都婚俗（Huadu's Marriage Customs）

现在很多节都被忽略了，但对于迎亲环节，花都人还是十分的尊崇。迎亲时，男家要请一对"好命"夫妇，为新郎采花（多为榕树或龙眼，寓意多子；或请父母双全、多兄弟的"案兄弟"负责采花）嵌床，然后进新房铺床，一边铺一边唱吉利歌，如"第一铺床铺席先，第二铺床铺草毡，第三铺床挂蚊帐，第四铺床铺花被"，花被中间有棵莲子树，莲珠莲子二十齐等，并在床的四角各摆一个母子相连的芋头（寓意多子），再放甜酒、生姜、红鸡蛋、谷、米、黄豆（寓意有种子）和铜盆鞋子（寓意同偕白首）、碗筷剪刀尺子（寓意有吃有穿）等物。接着一帮"案兄弟"给新郎穿长袍马褂，胸背挂绣球，戴上两边一对金花的礼帽，叫做"簪花挂灯"。女家则请来"好命婆"为新娘梳头，只梳三遍，边梳边吟唱"一梳梳到尾，二梳白发齐眉，三梳儿孙满地"等吉利话。临出门前，新娘的母亲交给新娘一把用红头绳将99个铜钱编织成的"钱剑"。

新娘在新婚后的第三天回娘家，男家要准备烧猪（俗称"金猪"）、圆子，一起带到女方家中。新娘回门，新郎送9条有头有尾的糖蔗，谓之甜甜蜜蜜、白头到老；女方家则必须将糖蔗的头、尾都裁下，回赠给女婿。

Many customs have been omitted, but the convention of escorting the bride is still very important for people of Huadu. During the escorting, the groom's family should invite a pair of couple who live a happy life to pick "flowers" (the flowers can be banyans or longans meaning more children. Or the family can invite the "brothers" who have biological brothers and alive parents to pick the flowers) and prepare the bed, then go to the bridal chamber to make the bed singing the auspicious song, such as "the first is matting, the second is straw, the third is mosquito net, the forth is colorful quilt, there is a lotus tree in the middle of the quilt, lotus bead and lotus seed are all together" etc. then put a connected-double—taros on the four corners of the bed (which has a meaning of more children), then put the rum, ginger, red egg, grain, rice, soybean (which has a meaning of seed), cooper basin and shoes (which has a forever love and marriage), bowls and chopsticks, scissor and ruler (which has a meaning of adequate food and drinks) and other items on the corners. Then the "brothers" help the bridegroom wear the long robe, the silk-made ball and the special hat with golden flower on each side, which is called "flower and lantern hanging". The bride's family should invite the "lucky woman" to comb the bride' hair for three times singing "the first comb should go to the end, the second comb means that the couple lives a happy and long-lasting live, the third comb means that they have many children" etc. Before going out, the bride's mother will give the bride a "Money sword", which is made of 99 copper cash strung by one red line.

花都节庆、习俗
FESTIVALS AND CONVENTIONS IN HUADU

The bride goes back three days after her marriage, and the groom's family should prepare the roasted pig (commonly known as the "Golden Pig") and dumplings to the bride's home. The groom should give 9 complete sugarcanes to the bride to looking forward to the sweetness and long-lasting marriage; the bride family should cut the top and the end of the sugarcane and give them back to the groom.

花都庆寿习俗（Birthday celebration convention in Huadu）

花都人过生日，并非年年都过，尤其是老辈人。花都人"做生日"，有"小生日"和"大生日"之分。

花都人一般在30岁前不做生日，否则会折寿。故小孩生日那天，父母只煮两个鸡蛋为贺，年纪稍长，完全不提。已婚男子头一次生日，是在30虚岁那年，由岳父母为女婿做生日，岳父母送来的礼物中，必须有一只刚会啼叫的生鸡（公鸡），称之为"啼"，希望女婿从此好运、发迹，此鸡不能杀，要一直养着。当天清晨，妻子必须为寿星煲鸡蛋糖水，寓意以后日子过得甜蜜、完美，寿星必须喝生日糖水。等到"男完女结"（子女全部结婚），儿女们为孝心，重提父母生日。每年备些酒肉、礼品祝贺父母诞辰，统称"小生日"，一般是家庭小贺。等到父母花甲之年（60岁）或花甲之后整十年，谓之"大生日"。60岁称"花甲"，是下寿；70岁称"古稀"，为中寿；80岁称"耄"，为上寿；90岁称"耋"，为高寿；100岁称"期颐"，是满寿。

另外，花都人过生日还男女有别，一般来说，花都男子喜欢做齐头（60、70、80），花都女子则有做"出一"（61、71、81）的习俗。

People of Huadu don't celebrate their birthday every year, which can be seen on the aged. There are two types of birthday celebration, one is "small birthday", and the other is "big birthday".

People don't celebrate the birthday before the age of 30, otherwise the life span will be shortened. So on the birthday of the children, the parents only cook two eggs for congratulation. They almost ignore the birthday celebration when the children grow up. The first birthday of the married man in the nominal age of 30, parents-in-law celebrate the birthday. Among the gifts given by the parents-in-law, one rooster which begins to crow is a must, which is called "Crow" and stand for the wish that the son-in-law has a good fortune. The rooster mustn't be killed until its natural death. The same morning, the wife should cook egg-sugar water for the husband, which means that they live a sweet and perfect life in the future. The husband must drink the egg-sugar water. Until the "completion of marriage of boys and girls" (all daughters and sons get married), the children will want to celebrate the birthday of their parents for filial piety. The children prepare some food and birthday gifts to congratulate the parent, which is called "small birthday" and celebrated among family member. Until their sixties or the whole ten years after the sixties, the celebration can be called "big birthday". "Sixties" is called "Huajia" and is Xiashou; "Seventies" is called "Guxi" and is Zhongshou; eighties is called "Septuagenarian" and is Shangshou; nineties is called "Octogenarian" and is Gaoshou; one hundred years old is called "Qiyi" and is Manshou.

In addition, there is a difference between the male and the female birthday celebrations. The men of Huade celebrate the birthdays at the ages of 60, 70 and 80 while the women celebrate the birthdays at the ages of 61, 71 and 81.

IMPRESSION OF HUADU 印象花都

❸ 看花都人的信仰大有乾坤
Rich history of faith of people in Huadu

观音诞（Date of birth of Avalokitesvara）

观音诞，花都人有供奉观音菩萨的习惯。观音诞一年有四次，农历二月十九日是观音生日，六月十九日是出家受戒日，九月十九日是成道登仙日，十一月十九日是入海为水神日，所以又称"南海观音菩萨"。

People of Huadu worship the Avalokitesvara. There are four births of Avalokitesvara, the birthday is on the nineteenth day of second lunar month. On the nineteenth day of sixth lunar month, she becomes the nun and learns the dharma, on the nineteenth day of ninth lunar month, she becomes the immortal, on the nineteenth day of eleventh lunar month, she is in the sea and that day is called "date of Arethusa", she is also called "Avalokitesvara of the South Sea".

花都盘古王诞（Birthday Celebration of King of Pan Gu in Huadu）

从清代嘉庆二十四年初重建狮岭盘古王庙以来，民间就把每年农历八月十二日定为盘古王诞，由十二日至十五日一连四天举行庆祝活动。时值夏收夏种之后，民间农闲，前来庆诞的群众人山人海，香火鼎盛，炮竹不绝，一连四天上演大戏、抢炮、闹花灯、舞狮，小商贩也进场摆卖，有香烛、食品、饮料、吉祥物、转运风车等出售，场面热闹非常。盘古王山的盘古王诞庆祝活动，在文化大革命期间停止进行，文革后逐步恢复，1985年盘古王庙和其他古迹、道路环境由民间集资修整后，1986年民间组织——盘古王理事会正式发动群众，重开庆诞，观众达6万人。以后每年庆诞日都有3万多人。区内部分村庄的狮子队参与庆诞。天刚亮，各地狮队锣鼓喧天，舞狮进场后，各狮队在盘古王庙前大广场一字摆开，同时起舞，庆诞活动达到高潮。近几年又恢复了演戏活动，一连四天下午、晚上演大戏（粤剧）或有歌舞表演。

花都节庆、习俗
FESTIVALS AND CONVENTIONS IN HUADU

Since the reconstruction of Temple of King of Pan Gu of Shiling at the beginning of 24th year of Emperor Jiaqing in Qing Dynasty, the folk select the 12th day of eighth lunar month each year as the birthday of King Pan Gu. There is four-day celebration from the 12th day to 15th day of lunar month. It is just after the summer harvest and summer planting, the unoccupied folk people, tens of thousands of villagers rush here from far and wide. There are much incense and candle, and continuous fireworks. Full-scale drama, gun jumping, lantern show and lion dancing last for four days. Small traders also do business in the temple. Candles, food, beverages, mascot and fortune-change windmills are sold. There are so many people that the temple is very crowded. The celebration of date of the birth of King Pan Gu stopped during the Great Cultural Revolution and restored gradually after the Great Cultural Revolution. In 1985, the temple, other historic sites and the road environments are restored and maintained through civic financing. In 1986, the civil organization--the Council of King Pan Gu officially mobilizes the masses to restart the celebration of the date of

birth of King Pan Gu, there are around 60,000 visitors on the 12th of eighth lunar month. After that, there are around 30,000 visitors in the birthday celebration each year. The lion-dancing teams of part villages in the district participate in the celebration. At dawn, the lion-dancing teams beat the gongs and drums. After the lion-dancing team enter into the arena, each team should line up and dance at the same time, which achieves the climax of celebration. The play is also restored in recent years. The full-scale drama (Cantonese opera) or dancing and singing performance lasts for four days, people can enjoy them in the afternoon or in the evening.

IMPRESSION OF HUADU 印象花都

"花果飘香，物候丰盛是花都。"

炭步芋头，"浮出水面"（Taro of Tanbu, "emerge from the water"）

炭步镇文岗村出产的炭步芋头有很长的历史了，乃至乾隆皇帝也成为炭步芋头的广告"代言人"。传说清朝乾隆微服游江南，有一次到达炭步圩，他在店里吃饭时问店主："有什么好吃的家乡菜？"店主介绍了一道文岗芋头焖扣肉，乾隆皇吃后大加称赞。他回到北京后，下旨花县知府每年进贡炭步文岗芋头，由此炭步芋头闻名天下。选购文岗香芋最好是到产地，只有文岗村种出的香芋味道最好，既松化可口又香气扑鼻。购买的时候，有一个小妙招，可以防止买到"假冒芋头"，只要将芋头放进水桶，浮出水面的才是真货，因为其他地方产的芋头都沉到水底去了。

"Taro of Tanbu" produced in Wengang Village of Tanbu Town has a long history and Emperor Qianlong even is the spokesman of it. It is said that during the private tour of South China in Qing Dynasty, one time King Qianlong arrived at Tanbu Dyke, he asked the owner in the restaurant, "Is there any delicious food?" the owner introduced the "Wenggang Taro&Pork". King Qianlong praised it highly after tasting. When he returned to Beijing, he decreed that the Prefect of Huaxian County pay tribute of Tanbu Wengang taro to the king, thus "Tanbu Taro" is famous in the whole nation. The best "Wengang Taro" should be purchased from its production place and the best flavor only can be tasted from the Wenggang taro, which is very delicious and nice-smelling. One tip can prevent you to buy false taro. You just put the taro in the water and if the taro emerges from the water, it is the original Wengang taro because other taros produced in other places sink to the bottom.

京塘莲藕，堪比鹿茸（The lotus of Jintang, as precious as pilose antler）

天下莲藕很多，但是仅有京塘盛产的莲藕才能配得上"京塘莲藕"这个"名号"。花东镇的京塘面积有80亩，莲藕纯天然生长，不用人工播种，每年秋末、冬初任人采挖。不需栽种，即取之不尽，用之不竭。京塘莲藕体型比普通莲藕细而且长，1.5～2米长，重1.5～2公斤，经过煲或炖，松化香醇。其淀粉含量13.28%，铁含量6.2%，锌含量3.8%，有较高的营养和药用价值，被人美称为"植物鹿茸"。

花都物候特产
SPECIALTY IN HUADU

"Smell the tragrance of fruit and flower everywhere, the rich resources is gathered in Huadu"

There are many kinds lotus in the world, but only the lotus from that one poor deserve the title of "Jingtang Lotus". The lotus of Jintang of Huadong Town grows naturally with an area of 80 Mu. People can get the lotus easily at the end of autumn and the beginning of winter. People can get the lotus continuously without planting it. The lotus of Jintang is longer than the normal lotus with a length of 1.5 m to 2 m and weight of 1.5 to 2 kilograms. The lotus taste delicious after stewing or braising. The lotus has a high nutritional and medicinal value with starch content of 13.28%, iron content of 6.2% and zinc content of 6.2% and called "Pilose Antler in Plant".

杨荷荔枝，百年历史（Yanghe leechee with a long history ）

"杨荷荔枝"在花都已有上百年的种植历史，种植面积有五六千亩，品种有桂味、糯米枝、槐枝，以老树槐枝为主。新鲜的荔枝每年6月底至7月初上市。鲜食、干制皆宜。槐枝肉乳白，软滑多汁，味甜带酸，核大而长，偶有小核。

"Yanghe leechee" in Huadu has been planted for hundreds of years with a planting area of five thousand to six thousand Mu. There are different kinds such as cinnamon-taste Litchi, nuomizhi and huaizhi, among which the old huaizhi is the main kind. The season of Fresh leechee comes across the June or July of each year. The leechee can be tasted directly or can be dried. The pulp is very delicious with much juice. It tastes a little bit with big and long kernel, sometimes the kernel is small.

李溪石峡龙眼，浓香爽脆（Shixia longan of Lixi, fragrant and crisp ）

李溪的"石峡龙眼"种植规模达2500亩，每年的7月底至8月初采摘，最大的特点是核小、皮薄、肉厚而爽脆，味清甜而浓香，每100克果肉含维生素C65.85～74.47毫克，营养丰富，具有壮阳益气、养血安神、润肤美容等功效，可辅助治疗贫血、失眠、神经衰弱及病后、产后身体虚弱等症。

"Shixia longan" of Lixi is planted with an scale of 2500 Mu. Its harvest period comes from the end of July to the beginning of August. The longan of Lixi has an obvious feature: small pit and fleshy and crisp pulp, it tastes sweet and smells fragrant. The longan is rich in nutrition with　C65.85-74.47 mg of vitamin per 100 pulp. It is helpful to kidney and spirit, nourish the blood and tranquilization, moisten the skin etc. It is also helpful to treat the anemia, insomnia, neurasthenia and poor health after disease or childbirth etc.

IMPRESSION OF HUADU 印象花都

"花果飘香，物候丰盛是花都。"

莘塘红蜜杨桃（Hongmi carambola of Shentang）

红蜜杨桃基地坐落在无环境污染的农田中，种植面积约100亩，每年的五六月和元旦前后采收。红蜜杨桃色泽鲜艳，金黄带微红，具有独特的蜂蜜型风味，果实肥厚，汁多渣少，肉质爽脆可口，它富含维生素及可溶性糖分，有清咽润喉、化痰止咳、生津止渴、醒酒解毒的作用。

"Hongmi Carambola Base" is located in the farmland without environmental pollution, covering a planting area of 100 Mu. The harvest period comes in May, June and around the New Year. Hongmi carambola has a bright color--golden red and a unique honey-type taste. It is a kind of fleshy fruit with much juice and less pit. Its pulp tastes crisp and sweet. The carambola is rich in vitamins and soluble sugars. It is helpful to clear and nourish the throat, eliminate the phlegm and stop cough, help produce saliva and slake thirst and sober up.

莘田二村萝卜，皮薄肉嫩
(Turnip of Zitian No.2 Village with thin surface and tender flesh)

莘田二村萝卜分"早白"、"迟花白"两种，生长于流溪河冲积而成的沙质土壤中。其皮色光洁、白净，个头大、粗纤维少、爽脆清甜、口感好且皮薄肉嫩，生吃、熟吃皆可。

Turnip of Zitian No.2 Village can be divided into two types: "zao bai" and "chihuabai". The turnip grows in the sandy soil formed by the impact of Liuxi Lake. Its peel is very smooth and the color is full white with less crude fiber and big shape. It is big and taste crisp, delicious with thin peel and tender pulp. The turnip can be ate directly or cooked to eat.

花都物候特产
SPECIALTY IN HUADU

"Smell the tragrance of fruit and flower everywhere, the rich resources is gathered in Huadu"

放养走地"狮前鸡"(Natural "Shiqian chicken")

花东镇的狮前村，其周边都是国家的生态公益林，山上的树木维系着九龙潭水库的水源。山上有很多野鸡，还有不少珍贵的鸟类，也有黄猄、野猪等动物。该村放养的鸡非常有名，很多人慕名前来品尝"狮前鸡"。

"Shiqian Village" in Huadong Town is surrounded by national ecological forest, trees on which is helpful to keep the water of Jiulongtan Reservoir clean. There are lots of pheasant, precious birds, muflone and boar and other kinds of animal. The natural chicken is so famous that attract many visitors for tasting the "Shiqian chicken".

梯面山水豆腐花
(Cold beancurd jelly made from the water of Timian Mountain)

梯面镇出产的豆腐花采用自然纯净的山泉水和优质黄豆、蜂蜜制作，豆腐花细腻滑嫩，香甜可口，洁净卫生。

The beancurd jelly of Timian Town is made from natural and pure mountain spring water, high-quality soybean and honey. The jelly taste delicate, smooth and very delicious.

IMPRESSION OF HUADU 印象花都

"花果飘香,物候丰盛是花都。"

花都咸水角（Salted dumpling of Huadu）

花都区农村在元宵节有人做咸水角吃。粘米粉和糯米粉混合,用水搅和,搓成皮,皮中放咸馅——馅由鲜肉丁、火腿丁、虾米混和,葱、蒜、韭、姜、蚝油、酱油等调味,再将皮对摺裹起成为角,这种咸水角称为"五味元宵",寓意聪明（葱）、会算（蒜）、长久（韭）、向上（姜）。做成的咸水角用油炸成微黄色,吃起来很可口。

People also make salted dumpling during the Lantern Festival. The wrapper of salted dumpling is made of rice flour and sweet rice flour, the fillings of which is made by fresh diced meat, diced ham and small shrimp mixed with shallot, garlic, leek, ginger, oyster sauce and soy sauce. After staffed with fillings, the wrapper should be folded into the angle. The salted dumpling is called "Five-kind filling dumpling", meaning wise (shallot), sophistication (garlic), long-lasting (leek) and progress (ginger). Fry the dumplings and taste very delicious.

花都臭皮醋（Choupi soup of Huadu）

"臭皮醋"这一方物实在是花都仅有,别无分号。儿媳怀孕期间,婆婆会用土法浸制"臭皮醋"（味臭食香）给儿媳饮食。"臭皮醋"的制作方法很简单,用瓦埕盛凉开水,放入炒米、姜、锅巴等,将瓦埕密封,放于开阳处晾晒三个月即可煮食。

"Choupi soup" can be found only in Huadu. During the pregnancy, the pregnant woman will eat it (smell rotten but taste good) cooked by the mother-in-law according to the old convention. It is quiet easy to make it. Put the water, parched rice, ginger and crispy rice etc together in the jar, then seal the jar. It is edible after three months later in the sunshine.

花都物候特产

"Smell the tragrance of fruit and flower everywhere, the rich resources is gathered in Huadu"

石湖莲藕（Lotus root in Shihu Village）

藕生于污泥而一尘不染，中通外直，不枝不蔓，自古就深受人们的喜爱。炭步石湖莲藕因其含有较高氨基酸、蔗糖、葡萄糖、钙、磷、铁及多种维生素而广受欢迎。石湖莲藕采用农家肥作为种植莲藕的基肥，种出的莲藕特别绵香。石湖莲藕种植面积不大，产量不多，但一到上市时间，众多采购商纷纷从各地而来，莲藕被抢购一空。

石湖莲藕还具有较高的药用价值，生食能清热润肺、凉血行瘀；熟吃可健脾开胃、止泻固精。老人常吃莲藕，可以调中开胃、益血补髓、安神健脑、延年益寿。

Lotus root rises unsullied from mud, symmetrically perfect; hence, it has been popular with people from ancient times. Lotus root produced in Shihu Village of Tanbu Town enjoys great popularity for the high content of protein amino acids, sucrose, glucose, calcium, phosphorus, iron and other vitamins. Fertilized by farmyard manure as base fertilizer, lotus root planted in Shihu Village is specially fragrant and sweet. Although the cultivated area is not large and volume of output is not high, once comes into the market, purchasers in large quantity come from all over one after another and lotus root is sold out.

In addition, Lotus root in Shihu Village possesses relatively high medical values that the uncooked ones are effective for clearing heat, moistening lungs, cooling blood and dispersing blood stasis while cooked ones are effective for strengthening spleen, stimulating appetite, antidiarrhea and securing essence. As for the aged, frequently eating lotus root could regulate the center, stimulate appetite, benefit blood, add marrow, calm the nerves and strengthen brain, with the effect of prolonging life.

生晒霸王花干（Suncured and Dried Pitaya Flower）

生晒霸王花干是天然生晒菜干，选取上等的霸王花晒干而成。最常的吃法为霸王花干煲猪肺。其主要产地在炭步镇骆村。

The natural suncured and dried pitaya flower is made from excellent Pitaya Flower which are dried in the sun. The most common cooking approach is pitaya flower boiled with pork lung, which mainly produced in Luocun Village, Tanbu Town.

洪秀全像
STATUE OF HONG XIUQUAN

新华街篇
XINHUA STREET
花都全攻略
OVERALL GUIDE FOR TRAVEL IN HUADU

新华街是花都区的政治、经济、文化中心，也是"天王"洪秀全出生、学习、研究带领革命的地方。探访伟人故居，您就像在经历一场心灵的洗礼，故事似乎埋藏在每一块砖每一片瓦中，只要耐心"温故而知新"，便能获得更多的心灵体悟。

"新华"是一个英雄辈出之所在。早在清朝年间，这块沃土就孕育了太平天国农民革命领袖洪秀全、洪仁玕、冯云山等英雄人物，他们创立"拜上帝会"，开展反封建专制、推翻清皇朝的革命斗争，谱写了近代农民革命史光辉的一页；辛亥革命时期，十八位新华籍的志士仁人在反封建的革命斗争中壮烈牺牲，被载入黄花岗七十二烈士名册。

Xinhua Street is the political, economical and cultural center of Huadu District, as well as the birthplace and the place where Hong Xiuquan learnt and studied the revolution. Visiting the former home of this great man, you are just like going through a spiritual baptism. Their stories seem to be buried in every brick and every tile. You can gain more spiritual experience through the way of "looking for the old so as to learn the new" patiently.

"Xinhua" is a place giving birth to a multitude of heroes. As early as Qing Dynasty, the leaders of revolution of peasants of Taiping Kingdom appeared on the fertile soil like Hong Xiuquan, Hong Renxuan and Feng Yunshan etc, who created the "the Congregation of the Worshippers", carried out the revolutionary struggle to oppose the feudal autocracy and overthrow the Qing Dynasty, writing a magnificent page in the revolution history of peasants of recent history; during the revolution of 1911, eighteen kind and upright men born in Xinhua sacrificed bravely and gloriously in the anti-feudal revolutionary struggle, whose name is wrote in the rooster of 72 martyrs of Huanghuagang.

数字看新华
FIGURES OF XINHUA

全街总面积112平方千米，其中城区面积111.03平方千米。
The whole street covers a total area of 112 square kilometers, of which the urban area is around 111.03 square kilometers.

GUIDE TO HUADU
花都全攻略 新华街 / XINHUA STREET

洪秀全故居
FORMER RESIDENCE OF HONG XIUQUAN

　　洪秀全故居位于广州市花都区新华官禄布村。1841年1月1日，洪秀全在花县福源水村出生后不久，全家迁到这里定居。洪秀全曾在这里读书、教书、耕作并从事早期的反清活动。

　　现在我们看到的洪秀全故居是1961年在旧址上复原的，同时复原的还有洪秀全青少年时期读书和教书的私塾"书房阁"。现在还保留有洪秀全手植的一棵龙眼树、洪秀全高祖洪英纶夫妇的画像、洪仁玕故居遗址和民房墙基等遗物。三块牌匾"广东省花县洪秀全纪念馆"、"洪秀全故居"、"书房阁"均为郭沫若先生题写。2005年被评为国家3A级旅游景区。

Former Residence of Hong Xiuquan is located in the Lubu Village, Xinhuagong, Huadu District of Guangzhou. On January 1st, 1841, the whole family moved here after his born in Fuyuanshui Village Huaxian County. Hong Xiuquan engaged in studying, teaching, farming and early rebelling Qing Dynasty activities here.

The present residence of Hong Xiuquan we see is restored on the basis of former residence in 1961 and the "Study Pavilion" where Hong Xiuquan studied in his adolescence and engaged in teaching is also restored at the same time. The longan tree planted by Hong Xiuquan, the portrait of the Hong Yingluns----great-great-grandfather of Hong Xiuquan, former residence of Hong Renxuan and bench-table of citizen house etc. Three plagues "Memorial Hall of Hong Xiuquan of Huaxian County of Guangdong Province", "Former Residence of Hong Xiuquan" and "Study Pavilion" are inscribed by Guo Moruo. In 2005, it was awarded the National Scenic Spot of AAA-Level.

★ 推荐看点
Recommended Viewing Focus

故居前面有一口鱼塘，塘中能看到远处丫髻山的倒影；附近有洪秀全族弟洪仁玕的故居遗址；村前有洪秀全读书、任教的书房阁和一棵由他亲手种植的有100多年历史的龙眼树。

There is a pool in front of the former residence, in which the Yaji Mountain is mirrored; the former residence of Hong Renxuan, the younger cousin of Hong Xiuquan is not far; the "Study Pavilion" where Hong Xiuquan studied and engaged in teaching is in front of the village; the longan tree planted by Hong Xiuquan has a long history of more than 100 years.

 游览攻略
Travel Guide

地址：花都区新华街官禄布村
电话：020-36861225　　门票：15元
公交线：花都公交车21路，花都洪秀全故居总站—花都豪利花园总站。
自驾线：机场高速—花都出口—迎宾大道—红棉大道约3千米处。

The address of Former Residence of Hong Xiuquan: Guanlubu Village, Xinhua Street, Huadu District
Tel:020-36861225　　　Tiket:15 Yuan
Bus route: Huadu No.21 bus, Station of Former Residence of Hong Xiuquan of Huadu–Station of Haoli garden of Huadu.
Self-driving route: Airport Expressway–Exit of Huadu–Yingbin Avenue–drive for 3 kilometers on Hongmian Avenue.

GUIDE TO HUADU

花都全攻略 新华街/XINHUA STREET

圆玄道观
YUEN YUEN TAOIST TEMPLE

　　道教教义有禁绝邪恶、引人向善的性质，所以受人崇敬。当下道家的精神也具有现实意义。金碧辉煌的圆玄道观处处笼罩着道家的飘逸和超然。置身其中，探访掩映的道观，聆听道教祖师骑着青驴、在百花深处成仙羽化的传说，使人摆脱了尘世的烦扰，忘却了人间的欲望。老子悠然的道家情怀抚慰了众多失意之人。问道圆玄，有人梦着而来，清醒回去；有人乘风而来，满载而归。2013年4月正式评为国家AAAA级旅游景区。

The Taoism is popular with people because of its creeds which restrict the evil and lead people to be kind. Today, the spirit of Taoism has practical significance. You can experience the elegance and aloofness everywhere in the magnificent Yuen Yuen Taoist temple. Standing in the temple, listening to the legend that the founder of Taoism becomes the immortal in the flowers riding a donkey, you will forget the earthly annoyance and ambition of the world. The leisurely feeling of Taoism of Lao-tzu comforts many frustrated people. During the visit of the Yuen Yuen Taoist Temple, people go back awake while he comes dreaming; some go back with fruitful results while comes with nothing.It was named the National AAAA Tourist Attraction in April 2013.

推荐看点
Recommended Viewing Focus

　　三清殿是圆玄道观的主体建筑物，采用古典天坛形式建造，天花绘太极图，寓意天大地圆光明无凝。大殿黄色琉璃盖顶，斗拱飞檐，彩绘绚丽。大殿中供奉玉清、上清、太清三圣铜像。

Palace of Three Immortals is the main architecture of Yuen Yuen Taoist Temple and is built in the form of classical Temple of Heaven, the diagram of the universe painted at the ceiling means the vast world, round earth and flawless light. The palace roof is covered with yellow tiles, the corbel bracket and overhanging eaves are also adopted with brilliant painting. The bronze statues of three immortals Yuqing, Shangqing and Taiqing are enshrined in the palace.

游览攻略
Travel Guide

地址：花都区新华街迎宾大道西38号(花都区九潭村)
电话：020-36861238　　　　门票：15元
公交线：花都公交车9路、70路，人和车站总站—花都万科天景花园总站，花都圆玄道观站。

Address: No.38 Yingbin Avenue West, Xinhua Street, Huadu District (Jiutan Village, Huadu District)
Tel: 020-36861238　　　　Ticket: 15 Yuan
Bus route: Huadu No.9 bus and No.70 bus, Renhe Station Termina-Huadu Wanke Tianjing Garden Terminal, Huadu Yuen Yuen Taoist Temple Station.

GUIDE TO HUADU
花都全攻略 新华街 / XINHUA STREET

资政大夫祠
ANCESTRAL TEMPLE OF ZIZHENG SENIOR OFFICIAL

资政大夫祠建于清代同治二年至三年（1863年），至今已有150年历史。实际上，我们现在看到的是资政大夫祠建筑群，是资政大夫祠、南山书院、亨之徐公祠和国碧公祠连建在一起的群体建筑。其中，南山书院是清兵部主事徐表正为其父徐时亮被赠奉直大夫而建的生祠；国碧公祠和亨之徐公祠则是徐姓的祠堂。

ncestral Temple of Zizheng Senior Official has a history of 150 years, built from the second year to third year (year 1863) of Emperor Tongzhi in Qing Dynasty. In fact, what we see now is "Ancient Architectural Complex of Ancestral Temple of Zizheng Senior Official", including Ancestral Temple of Zizheng Senior Official, Nanshan Classical Academy, Ancestral Temple of Heng Zhixu and Guobi Ancestral Temple. Among which, Nanshan Classical Academy is built by Xu Biaozheng, the Chief Master of the Ministry of War (in feudal China) to celebrate his father's promotion—being the Fengzhi Official (official name in ancient times); Ancestral Temple of Heng Zhixu and Guobi Ancestral Temple are the ancestral hall for the Xu family.

推荐看点
Recommended viewing focus

牌坊建于同治二年，四柱三间四楼，连州青石打制，面宽9.35米、高9.8米，明间设两层庑殿顶，上层檐下有石制的如意斗拱，正面悬刻"圣旨"，背刻"恩荣"及同治年间所颁的诏书。只有皇帝亲自下令"颁奖"的人物，祠堂牌坊才能镌刻上"圣旨"二字。

The memorial archway, built in the second year of Emperor Tongzhi in Qing Daynasty, is made of bluestone of Lianzhou. It is supported by four pillars, forming three rooms and four floors. The archway has a side width of 9.35 m and a height of 9.8 m. The outer room is designed to adopt two-layer hip roof, and the front side of Ruyi bracket made from good eaves-stone is inscribed "Imperial Edict", the back side is inscribed "Glory" and the imperial edicts issued in Year of Yongzhi. Only for the person who gets the award from the emperor personally, his memorial archway can be inscribed the two words "Imperial Edict".

游览攻略
Travel Guide

地址：资政大夫祠位于花都区新华街三华村的西面
电话：020-86903818　门票：15元
公交线：花都公交车9路（花都汽车站总站-花都九塘总站，在"花都资政大夫祠"站下车）
自驾线：机场高速—花都出口—迎宾大道—左转107国道2千米处。

Address: west side of Sanhua Village, Xinhua Street, Huadu District
Tel: 020-86903818　Ticket: 15 Yuan
Bus route: Huadu No.9 bus (Huadu Bus Terminal—Huadu Jiutang Terminal), get off at "Ancestral Temple of Zizheng Senior Official"
Self-driving: Airport Expressway—Exit of Huashan-Yingbin Avenue—turn left to the G107 and drive for 2 kilometers.

GUIDE TO HUADU
花都全攻略 新华街 / XINHUA STREET

石头记矿物园
SHITOUJI MINERAL PARK

石头记矿物园是世界上首座以矿物为概念所建的主题乐园，国家AAAA级旅游景区。"矿物园"主体为一扇形建筑物，地下一层包括时空隧道，影视厅，水晶、琥珀原生态区，石头禅室，古代生物化石区和奇石矿物陈列区。奇石陈列馆中将展示来自世界各地的奇宝玉石；水晶世界引领您神游古今，穿透星球的奥秘；梦幻水晶餐厅，让游客可以在众多高能量的水晶包围圈中就餐，品尝正宗的台湾美食。

矿物园内有一座很特别的"古代森林"，即木化石展区。数百株亿万年高龄的木化石"生长"其中，游客行走于"林间小径"，感受来自远古森林的气息，感叹大自然造物的神奇。

The first theme park built with mineral as the concept in the world. The National AAAA Tourist Attraction The main body of the "Mineral Park" is a fan-shape building. The ground floor includes Spatio-Temporal Tunnel, Movie Theatre, Original and Ecological Zone of Crystal and Amber and Stone Hall, Ancient Fossil Area and Rare Stone and Mineral Exhibition Area. The Rare Stone Exhibition Hall shows rare gem and jade; the Crystal World leads you to go through the past and present, experiencing the mystery of the planet; in the Fantasy Restaurant, the tourist can enjoy the authentic Taiwanese cuisine in the high-energy atmosphere surrounded by crystal. There is a very special "Ancient Forest" in the "Mineral Park"–Silicified Wood Exhibition Area. Hundreds of silicified woods, which are hundreds of millions of years old, are displayed in the park. When the visitors walk in the "forest trail", they can feel the mystery of ancient forest, being surprised at the mystery of the nature.

推荐看点
RECOMMENDED VIEWING FOCUS

石头记矿物园中有传国玉玺、夜明珠、满汉全席、海洋之梦等石头艺术文化经典作品，尤其是以各种主题为造型的"满汉全席"，以其形似和神似，吸引了最多眼球。

The Shitouji Mineral Park have various classical stone art such as Imperial Seal of Country (used in ancient time), Garnet, Manchu Han Imperial Feast and Dream of Ocean etc. in particular, "Manchu Han Imperial Feast" with a variety of theme as the style attracts many visitors because of its appearance similarity and spirit similarity.

GUIDE TO HUADU

花都全攻略 新华街 / XINHUA STREET

石头记矿物园
SHITOUJI MINERAL PARK

游览攻略
TRAVEL GUIDE

神秘宝藏开放时间：9:00-18:00(票价60元)
芝麻开门开放时间：9:00-18:00(入园免费)；
　　　　　　　　18:00-21:00(票价5元)

自驾线一：机场高速公路，可从太成出口(花都出口)下高速，沿迎宾大道直行到底(约15千米)，右转行驶300米，到达石头记矿物园。
自驾线二：经广清高速公路，在赤坭或新华出口下高速，经风神大道右转至红棉大道，直行至迎宾大道左转到底后，右转约行驶300米到达石头记矿物园。
公交线：可搭乘花都区9路公交车至石头记站；或搭乘花都区21路公交车至石头记站。

Opening time of Mysterious Treasure: 9:00-18:00 (ticket price is 60 Yuan)
Opening time of Open Sesame: 9:00-18:00 (free admission)
18:00-21:00 (ticket price is 5 Yuan)
Self-driving route 1: via the Airport Expressway, get off at the exit of Taicheng (exit of Huadu), drive straight to the end of the Yingbin Avenue (around 15 kilometers), turn right and drive for 300 meters to the Shitouji Mineral Park.
Self-driving route 2: via Guangqing Expressway, get off the expressway at exit of Chini or Xinhua, turn to the Hongmian Avenue via Fengshen Avenue, drive straight to Yingbin Avenue and turn left to the end, turn right and drive for 300 meter to arrive at the Shitouji Mineral Park.
Bus route: take the Huadu No.9 bus to Shitouji Station; or take the Huadu No.21 bus to Shitouji Station.

云峰翡翠梦想馆
YUNFENG JADEITE MUSEUM

  云峰翡翠梦想馆由云峰企业投资兴建，是广州市花都区政府打造花都珠宝旅游文化产业的重点项目之一。云峰翡翠梦想馆是以翡翠为主题，汇集原创艺术、新型多媒体及3D动漫视觉体验、高科技数码触碰互动、火山熔岩体验、翡翠文化知识介绍、翡翠高端精品展览、艺术家欧式咖啡厅、文创精品等丰富内容版块的文化创意乐园。以创新的艺术思维及高科技的硬件设施、国际化团队的动漫制作、个性鲜明的原创动漫形象、丰富多彩的主题活动，让您在快乐、惊喜的体验中领略翡翠艺术的迷人魅力。云峰翡翠梦想馆，一个具有高尚品位的翡翠与艺术博物馆，一个能带给您梦幻般体验的知性乐园！

  YunFeng Jadeite Museum, one of the key projects in the jewelry tourism and cultural industry established by the government of Huadu District, Guangzhou, is built and invested by Yunfeng Group. With the jadeite as the theme, it is a cultural and creative theme park including original art, new multimedia, visual experience of 3-D animation, digital touch interaction of high-tech, volcanic lava experience, knowledge introduction of jadeite culture, high-quality jadeite exhibition, European-style coffee house of artist and high-quality products of culture and creation etc. Innovative art thinking, high-tech hardware facilities, international team of animation production, distinctive and original cartoon image and colorful theme activities make you experience the charm of jadeite art in happiness and surprise. YunFeng Jadeite Museum is a museum of jadeite and art with refined taste and an intellectual paradise bringing you fantastic experience.

GUIDE TO HUADU
花都全攻略 新华街 / XINHUA STREET

👁 云峰翡翠梦想馆
YunFeng Jadeite Museum

⭐ **推荐看点**
Recommended Viewing Focus

　　3D动漫视觉体验、高科技数码触碰互动、火山熔岩体验、翡翠高端精品展览、艺术家欧式咖啡厅、文创精品。

Visual experience of 3-D animation, digital touch interaction of high-tech, volcanic lava experience, high-quality jadeite exhibition, European-style coffee house of artist, high-quality products of culture and creation.

游览攻略
TRAVEL GUIDE

地址：花都区镜湖大道2号
公交线：在广州北站或新世纪广场乘坐机场快线9号线，至云峰大酒店站，步行至目的地。
自驾线：机场高速—花都出口—迎宾大道—镜湖大道北—云峰商业大厦旁。

Address: No.2 Jinghu Avenue, Huadu District
Bus route: Take the Airport Express Line 9 at Guangzhou North Railway Station or New Century Plaza to Grand Peak Hotel, and walk to the destination.
Self-driving route: Airport Expressway—Exit of Huadu-Yingbin Avenue—Jinghu Avenu North—beside the Yunfeng Commercial Building.

GUIDE TO HUADU
花都全攻略　新华街／XINHUA STREET

👁 花都汽车产业基地
HUADU MOTOR INDUSTRY BASE

　　花都汽车产业基地位于花都区风神大道的花都汽车工业城，是由东风汽车公司、台湾裕隆汽车制造股份有限公司、广州京安云豹汽车有限公司共同组建，由东风汽车公司控股的大型汽车企业。该汽车城将集汽车生产、贸易与营销、汽车仓储与现代物流、展示与博览、汽车科普教育与信息服务、文化与旅游等多功能于一体的现代化综合性汽车城。该汽车城总占面积为15平方千米，是中国南方地区汽车贸易的核心市场。

Located in the Fengshen Avenue of Huadu, the Motor Industry City of Huadu is actually large-scale automobile company held by Dongfeng Motor Corporation and established by Dongfeng Motor Corporation, Yulon Motor Co., Ltd and Guangzhou Jingan Yunmao Motor Co., Ltd etc. It is the modern and comprehensive motor city integrating the motor production, trade and marketing, motor warehousing and modern logistics, exhibition and Expo, automotive science education and information services and culture and tourism etc. Covering a total area of 15 square kilometers, it is the core automotive trading market in the south of China.

 游览攻略
TRAVEL GUIDE

地址：花都区新华街风神大道1号
公交线：在花都客运站乘坐花都19路永高班车（或花都19路花都港班车、花都19路空调、花都27路），在广东省岭南工商第一技师学院站下车，步行至目的地。
自驾线：机场高速花都入口—迎宾大道—镜湖大道北—雅瑶中路—X280—南浦大桥—X283—东风大道—花都汽车城。

Address: No.1 Fengshen Avenue, Xinhua Street, Huadu District
Bus Route: Take the Huadu No.19 bus–Yonggao shuttle at the Huadu Passenger Station (or Huadugang shuttle or Huadu No.19 air-conditioned bus or Huadu No.27 bus) to the Guangdong Lingnan No.1 Technical Institute of Industry and Commerce, then and then walk to the destination.
Self-driving route: Entrance of Huadu of Airport Expressway–Yingbin Avenue–Jinghu Avenue West–Yayao Middle Road–X280–Nanpu Bridge–X283–Dongfeng Avenue–Huadu Motor City.

GUIDE TO HUADU
花都全攻略 新华街 / XINHUA STREET

花都好吃

花都圆仔
ROUND DUMPLING IN HUADU

花都的圆仔最有特色，是一种米粉做的小吃。搓成圆条状后再切成一小根一小根，放入瓜菜和肉类同煮，可以有很多种搭配，味道鲜美、可口，还可以当主食，一举两得。圆仔是花都人喜爱的一种小吃，很多家庭主妇都会做，但因为制作比较麻烦，所以现在大家都跑去酒楼吃。圆仔最特别之处是可以用很多瓜菜和肉类搭配，或煮汤、煮粥，或煎炒。

The round dumpling in Huadu is most distinctive and actually a kind of snack made of rice flour, which is kneaded into stripes and then cut into small piece, vegetables and meat should be added and cooked for a while. There are many kinds of ways to eat, it is very delicious and also can be the main food and very convenient. The round dumpling is very popular with people in Huadu, many is capable of making it. Due to the complex making method, people prefer to enjoy the round dumpling in the restaurant. The most particular aspect is that the round dumpling is edible together with many vegetables and meat being the soup, or porridge or frying etc.

花都烧骨粥
PORRIDGE WITH ROASTED BONE IN HUADU

花都烧骨粥是粥类中的佼佼者。花都烧骨粥之所以特别，是因为粥里用的烧骨都是现烧的，浓香无比之余，又不觉得油腻，刚烧出来的烧骨的香味渗透到浓稠的白粥当中，让人口齿留香。

The Porridge with Roasted Bone is the top-grade porridge among various kinds of porridges. The porridge is particular because the bone used in the porridge is freshly roasted with incomparable flavor and don't make you feel greasy. The newly cooked porridge with roasted bone is very delicious, when you taste it, the lingering fragrance will leave in your mouth.

Gourmet in Huadu

花都饮食风情街
Huadu Catering Gourmet Street

喜立登饮食风情街是一个集购物、饮食、休闲为一体的娱乐广场，分为中华名菜区、粤菜区、外国菜区、快餐美食区、娱乐休闲区共五大系列六大板块，距区政府仅800米，是花都东区新地标。

Xilideng Catering Gourmet Street is a comprehensive entertainment plaza integrating shopping, diet and relaxation and divided into five main series and six parts: Chinese famous dish area, Cantonese dish area, foreign dish area, fast-food and gourmet area as well as entertainment and relaxation area, only 800 meters away from the District Government, being the new landmark of Eastern Huadu.

食肆推荐Restaurant recommended：大宅门（Grand Mansion）

这里是众多食家极力推荐的地方。以美味时尚健康的干锅鸭头、干锅鱼、龙凤香锅为主营菜品，装潢环境挺不错。特色干锅鸭头老少皆宜，让人百吃不厌。点拉面还可观赏拉面表演。

必尝推荐：干锅鸭头、干锅虾、麻辣虾蟹、红酒雪梨。

It is highly recommended by the numerous food critics. The delicious and fashionable pot duck head, pot fish and Longfeng pot are the main dish. The restaurant is decorated gloriously. No matter the aged or the children can enjoy the pot duck head, which is so delicious that clients want to enjoy it again and again. If the clients order the pulled noodles, the pulled noodles performance is a gift.
Recommended delicacies: Pot Duck Head, Pot Shrimp, Spicy Shrimp and Crab, Poached Pears with Red Wine.

食肆推荐Restaurant recommended：祥记猪肚鸡（Xiang's Pork Tripe Stew Chicken）

主打猪肚包鸡，最小的中煲也足够4人份量。汤很鲜，加了牛奶所以非常香滑。胡椒味适中，冬天能暖胃。鸡是走地鸡，皮厚，口感润滑但不肥腻。周末人多，高峰期需要等位。

必尝推荐：猪肚包鸡锅。

One of the main dishes—Pork Tripe Stew Chicken, the smallest medium pot is enough for 4 clients. The soup is very delicious and smooth due to the milk. Pepper is added suitably in the dish, which is helpful to warm the stomach. The free range chicken used is tender but not greasy with thick skin. It is crowded during the weekend and you need to wait during the rush hours.
Recommended delicacies: Pork Tripe Stew Chicken Pot.

地址：百寿南路站，花都喜立登饮食风情街
交通：花都公交19路、22路
Address: Baishou South Road, Huadu Catering Gourmet Street
Bus route: Huadu No.19 bus, Huadu No.22 bus.

GUIDE TO HUADU

花都全攻略 新华街 / XINHUA STREET

 花都好吃

百利鸽王
BAILI PIGEON RESTAURANT

正宗的乳鸽专门店，出品只有乳鸽，主打的吊烧乳鸽和药膳乳鸽堪称经典！烧乳鸽肉中有汁，骨香酥脆。药膳乳鸽汤头鲜美且滋补，吃完鸽肉再喝汤，还可做火锅。大多时候需提前订座。

必尝推荐：鸽子、全鸽宴。

It is a genuine shop for eating the young pigeon. All dishes are related to the pigeon. Roasted Pigeon and Stewed Pigeon with Herb are classical dishes here! The roasted pigeon is very tasty and delicious. The soup of Stewed Pigeon with Herb is fresh and nutritive, which also can be used to cook the hotpot after finishing the pigeon. Usually you need to do the reservation in advance. Recommended delicacies: pigeon, all-pigeon feast.

 地址：紫薇路东（光明花园附近）　订座电话：13902395283
Address: East of Ziwei Road (near Huangming Garden)
Reservation Hotline: 13902395283

肥波烧鹅
ROASTED GOOSE OF FEIBO

很多广州人吃完返寻味，吃完意犹未尽，经常打包回广州。

必尝推荐：烧鹅。

Many people from Guangzhou ruminates the delicious goose and want to taste and they often pack the goose and take back to Guangzhou.
Recommended delicacies: roasted goose.

 总店地址：秀全路60号之7铺　　分店地址：曙光路30号
Address: Store 7, No.60 Xiuquan Road
Branches Address: NO.30 Shuguang Road.

Gourmet in Huadu

金骏酒家
Jinjun Restaurant

秉承做地道农家菜的特色，自开业以来就以优质的出品、大众化的价格赢得广州食家们的口碑。标准的粤菜口味，菜式颇具特色，其特色花都圆仔是花都远近驰名的。
必尝推荐：圆仔、电饭煲鸡。

With authentic farm food as its feature, it is popular with customers through delicious dish and cheap price since its opening. Its dish is quiet distinctive and the Cantonese dish is authentic. Huadu Round Dumpling is quiet famous from far and wide.
 Recommended delicacies: round dumpling, Chicken Cooked in Electric Cooker.

总店地址：宝华路东五小侧（骏威广场对面）
分店一地址：新华路52号，洪秀全纪念馆侧
分店二地址：商业大道东与迎宾路交界
Headquarter Address: Besides No.5 Primary School, Baohua Road East (opposite to Junwei Plaza)
Branch store 1 Address: No.52 Xinhua Road, besides Memorial Hall of Hong Xiuquan
Branch store 2 Address: at the intersection of Shangye Avenue East and Yingbin Road

---

丽雅人家
Liya Restaurant

多年专营茶市而闻名，每天能供应160多种点心，可谓是花都点心之最，口碑极佳。朋友相聚聊聊天，或想找个悠闲地方轻松谈谈生意，客人都会选择来这里。

必尝推荐：味皇蒸排骨、焦糖炖乌鸡蛋、铜盆蒸螺、丽雅斋粉果。

It is a famous tea café for many years, the daily supply of 160 kinds of dessert tops Huadu with a good reputation. It is a good choice for gathering among friends or business negotiation!
Recommended delicacies: Steamed Pork Ribs of Weihuang, PSilkie Egg Stewed with Caramel, Chicken Steamed in Copper Pot, Steamed Dumpling with Pork of Liya Restaurant.

地址：宝华路五小旁（骏威广场对面）
Address:Besides No.5 Primary School, Baohua Road (opposite to Junwei Plaza)

GUIDE TO HUADU
花都全攻略 新华街 / XINHUA STREET

花都好吃

椰林海鲜码头
SEAFOOD WHARF IN COCONUT GROVE

"只会做海鲜"的椰林海鲜码头,创造了花都几乎日日"放号"一百多位的酒楼奇迹。独创的海鲜"传送带"既提升效率又让客人看个新鲜,是一大特色,所聚集的海鲜品种也堪称"花都之最"。装潢颇具海滨风情,出品量身定制,口味独到,让人流连忘返。

必尝推荐:粉丝扇贝、炭烧生蚝、果汁煮花蟹、冰镇鳝片。

The "Wharf" only cooks the "seafood", creating a miracle of "allocating number" for over 100 in Huadu. The unique "conveyor belt" of seafood not only improves the efficiency, but also attract the customers, being the major feature of the restaurant. The variety of seafood here tops Huadu. It is decorated in seascape-style. The dish here can be cooked according to the customer's requirements with a so unique flavor that customers forget to leave.
Recommended delicacies: Scallops & Vermicelli, Charcoal Roasted Oyster, Boiled Spotted Crab with Juice, Chilled Eel Slices.

 地址:三东大道(人民公园对面)
Address: Sandong Avenue (opposite to People's Park)

湖景鱼翅海鲜城
HUJING SHARK FIN SEAFOOD RESTAURANT

毗邻秀全公园,装修豪华,设有大型海鲜自选市场,首创花都开放式明厨酒楼,以经营粤港新派菜式、鲍、燕、翅及高档海鲜为主,是花都地区首屈一指的粤菜餐饮品牌,多次承接政府级重要宴会,荣获多项殊荣。

必尝推荐:椰汁官燕、日本品鲍、浓汤鸡煲翅。

Located near the Xiuquan Park, it is decorated luxuriously, including a large seafood supermarket, being the first restaurant whose kitchen is open to the customer in Huadu with the new Cantonese-Hong Kong dish, abalone, edible bird's nest, shark's fin and other high-end seafood. It is the first-grade Cantonese cuisine brand in Huadu, undertaking the important feast with a government-level for many times. It is granted various awards for many times.
Recommended delicacies: Bird's Nest with Coconut Juice, Abalone from Japan, Shark Fin Soup Pot with Chicken.

 地址:滨湖大道36号湖畔花园广场
Address: Hupai Square Garden, No.36 Binhu Avenue

Gourmet in Huadu

食养坊
SHIYANG RESTAURANT

以药膳滋补炖汤、养生药膳菜和农家菜为主的全国绿色餐饮连锁企业。食材新鲜，菜式丰富且精致，粤式口味正宗。炖汤滋补营养，清淡味香。吃腻了大鱼大肉来此换换口味挺好。

必尝推荐：各式药膳炖汤、海鲜、药膳粤菜。

It is a national green food chain with Chinese herb soup, healthy Chinese herb dish and farm food as the main dish. The food material is fresh. It has various and delicate dishes. The Cantonese-style dish tastes very authentic. The stewed soup is full of nutrition and delicate with fragrance. It is a good choice to change eating style if you are tired of abundant fish and meat---rich food.
Recommended delicacies: a variety of stewed soup with Chinese herb, seafood, Cantonese cuisine with Chinese herb.

总店地址：丽雅直街3号　　订座电话：020-36989922
分店地址：茶园路13号（竣威广场后门）
Headquarter Address: No.3 Liyazhi Street　　Reservation hotline: 020-36989922
Branch store Address: No.13 Chayuan Road (back door of Junwei Plaza)

炳记美食
BING'S CUISINE

这里的肠粉得到客人翘首称赞，依照广东肠粉做法加以改进，肠粉又薄又脆，口感更好；品种非常丰富，肉类嫩滑，青菜脆口，美味独特。其他广式小吃与快餐也质量极佳。

必尝推荐：各式肠粉、鲜虾云吞、椰子炖乌鸡汤。

Steamed Vermicelli Roll is highly praised by the customer, whose cooking method is improved based on the method of steamed vermicelli roll of Guangdong. The roll is very thin and crisp and very delicious; there are variety of steamed vermicelli roll, the meat is tender and vegetable is crisp with a unique flavor. Other Cantonese-style snacks and fast food are also delicious.
Recommended delicacies: a variety of Steamed Vermicelli Roll, Wanton with Shrimp, Chicken Soup with Coconut.

总店地址：育才街建成巷9号
分店地址：天贵路88号
Headquarter Address: No.9 Jiancheng Lane, Yucai Street
Branch store Address: No.88 Tiangui Road

GUIDE TO HUADU
花都全攻略 新华街 / XINHUA STREET

花都好吃

华西面店
HUAXI NOODLE RESTAURANT

家庭作坊式的传统面店，新华路总店口碑坚固，环境简洁干净。云吞是招牌菜，绝对是自家手工制造，皮较厚不会煮烂，肉鲜味十足。面是碱水面，虽不如银丝面弹牙，但很有韧性。

必尝推荐：手工净云吞、手工蛋面。

The traditional noodle restaurant in the form of family workshop, the head store located in the Xinhua Road is very famous and clean with a simple decoration. Wonton is the famous dish and hand-made for certain. The wrapper is thick and not easy to be loose. The meat of the filling is very fresh. The flour is mixed with soda in the making. The noodle is full of elasticity, even though not as tasty as Yinsi Noodle.
Recommended delicacies: Hand-Made Net Wanton, Hand-Made Noodle with Egg.

总店地址：花都区新华路7号　　分店地址：新华街曙光路30号
Address: No.7 Xinhua Road, Huadu District
Branches Address: NO.30 Shuguang Road, Xinhua street.

永兴点心
YONGXING DESSERT

遍布花都的早餐连锁加盟店，主要卖各种点心，品种丰富，美味平价。糯米鸡馅料多且足量，糯米绵软入味；鲜肉包的肉汁鲜甜，猪肉新鲜爽口。奶茶味道也不错，性价比很高。

必尝推荐：糯米鸡、鲜肉包、芋丝糕。

It is a popular breakfast chain store in Huadu. There are a variety of delicious desserts here with cheap price. Sticky Rice in Lotus Leaf is tasty and soft with full filling; the Fresh Meat Package taste delicious and sweet and the meat used is fresh. Milky tea is jammy with cheap price.
Recommended delicacies: sticky rice in lotus leaf, fresh meat package, taro cake.

 总店地址：秀全大道64号
　　　　　花城店地址：花城路65号
Headquarter Address: No. 64 Xiuquan Avene
Huacheng Store Address: No.65 Huacheng Road

Gourmet in Huadu

黄师傅蛋挞
Egg Tart of Master Huang

黄师傅蛋挞有"花都蛋挞之王"的美称，几乎每次去都能看到一条长长的队伍。皮酥蛋香个大，性价比极高。菠萝包和老婆饼也很受欢迎，酥脆可口。

必尝推荐：蛋挞、菠萝包。

Its egg tart is called the "King of Tart in Huadu". There is always a long queue in front of the store. The egg tart is very delicious with crisp surface and is relatively big. You can enjoy it without high price. The pineapple
Recommended delicacies: egg tart, Pineapple Bread.

地址：新华路79号成功大厦107铺
Address: Shop 107, No.79 Xinhua Road, Chenggong Building

私厨"青砖屋"
Homely "Qingzhuan House"

没有店名，食客们为这家店起了"青砖屋"的代号。客流很旺，食材好、新鲜，饭任吃、汤任喝，很有家庭的味道。

必尝推荐：私厨家常菜。

There is no specific name. The diners call it "Qingzhuan House", which is very popular with people. The food material is very fresh. The rice and soup are ready for you all the time. You are just like in your home when dining here.
Recommended delicacies: Homely Dish.

地址：新华街紫薇路、百和家园附近
Address: Ziwei Road, Xinhua Street, near Baihe Home

满屋MX甜品
Manwu MX Sweets

花都一家经营多年的芝士蛋糕专门店，出品的芝士蛋糕芝士味非常浓郁，款式丰富，外表精巧。

必尝推荐：特浓芝士蛋糕、提拉米苏、芝士绿茶蛋糕。

It is a store only for cheese cake for many years in Huadu. The cheese cake is very delicious with rich fragrance. There is a variety of cheese cake with delicate appearance.
Recommended delicacies: Espresso cheese cake, tiramisu, cheese Cake with a flavor of green tea.

地址：花城路29号108-109铺(近新世纪酒店)
Address: Store 108 to 109, No.29 Huacheng Road (near New Central Hotel)

香草世界
VANILLA WORLD

花山镇篇
HUASHAN TOWN
花都全攻略 — OVERALL GUIDE FOR TRAVEL IN HUADU

花山镇原名花县。是原县府所在地。花山镇与"花山"有着莫大的关系,花山自清代花县城设立后,一直是县治的所在地,历经200多年;而且花山里还有一座名副其实的"花山"——菊花山,闻名遐迩,因出产异石,形如菊花,故得名"菊花山",所产的"菊花石"为广东名石。花山镇不仅石头漂亮,花山也是人杰地灵,客家山歌与铁山村灰塑艺术都是佼佼者,均列入了广东省非物质文化遗产名录。

Huashan Town formerly called "Huaxian County" which might have a close relationship with the "Huashan Mountain". The town is always the place of country jurisdiction since the establishment of Huaxian County in Qing Dynasty for over 200 years; there is a veritable "Huashan Mountain"----Chrysanthemum Mountain, which is known for its rare stones, which looks like chrysanthemum, so it is called "Chrysanthemum Mountain", the "Chrysanthemum Stone" is a kind of famous stone in Guangdong Province. Huashan Town is not only famous for its beautiful stones, but also for the outstanding people from Huashan Town. Hakka's folk song and ash art (outdoor decoration art in Guangdong Province) are renown in the world, and listed in the Nonmaterial Cultural Heritage of Guangdong Province.

数字看花山
FIGURES OF HUASHAN

花山镇位于花都区东部,紧邻广州白云国际机场,面积116.40平方千米,辖26个行政村和1个社区居委会。106国道、机场高速公路北延线、广乐高速、肇花高速、花都大道、三东大道和山前旅游大道等主干道在花山镇境内纵横交错,构成四通八达的交通网络。花山山前大道农业生态旅游片区规划已纳入区政府重点项目推进。

Located in the east part of Huadu District, adjacent to the Guangzhou Baiyun International Airport, Huashan Town covers an area of 116.40 square kilometers, having jurisdiction over 26 administrative villages and one community committees. Different main stems like G106, northern extension of Airport Expressway, Guangle highway, Zhaohua highway, Huadu Avenue, Sandong Avenue and the Shanqian Travel Avenue etc cross together in the Huashan Town, constituting a road transportation network extending in all directions. Avenue of Shanqian farm stay ecotourism tourism distict plan has been incorporated in the government key projects.

GUIDE TO HUADU
花都全攻略 花山镇 / HUASHAN TOWN

香草世界
VANILLA WORLD

香草世界是中国首家以种植香草为主题的公园，收集和引种世界各地著名香草，全力打造香草世界生态农业基地。这些香草中堪称世界著名美味蔬菜的极品有香罗勒、迷迭香、百里香、鼠尾草、薰衣草、香蜂草、柠檬香茅、洋甘菊、奥勒冈。

园区分为东方普罗旺斯、薰衣草王国、香薰园、玫瑰园、万国蔬果园、薰衣草木屋、婚纱摄影、农家博物馆、儿童手工课堂、美食点心坊、儿童科普教育园、香草食用研究基地等项目。

Vanilla World is the first theme park to grow vanilla in China, where collects and introduces the famous vanilla of the world, among which basil grand vert, rosemary, thyme, salvia, lavender, balm, lemongrass, chamomile and origanum are the delicious and famous vegetables in the world. The park is established with an object to be the world's ecological agriculture base of vanilla.

The park is divided into many parts such as Eastern Provence, Kingdom of Lavender, Lavender Garden, Rose Garden, Vegetable and Fruit Garden of World, Lavender Chalet, wedding photography, Museum of Peasant, Handcraft class of children, Cuisine and Snack Store, Children's science education park, edible vanilla research base.

RECOMMENDED VIEWING FOCUS

建议大家用"摸、闻、赏、尝"四字诀来参观游览香草世界。摸：用手轻轻在香草上带过；闻：放鼻中轻闻，满足味觉享受；赏：欣赏来自不同国家的香草，置身于香草海洋中感受香草文化；尝：品尝花卉、香草概念美食，全身心体验香草的内在魅力。

A suggestion for visiting the vanilla world: four-word formula "touch, smell, enjoy and taste". You can touch the vanilla gently; you can also smell it gently to experience the vanilla's aroma; you can enjoy vanilla from different countries, standing in the vanilla's ocean to enjoy the vanilla culture; you can taste the flower and the concept gourmet of vanilla, indulged in the inherent charm of the vanilla.

TRAVEL GUIDES

地址：花都区花山镇铁山村
电话：020-86788172   门票：100元
公交线：乘坐花都4路公交车到总站，再转乘摩托车至目的地。
自驾线1：机场高速—花都出口—迎宾大道—106国道—284县道（沿X284行驶2千米，右转）—据指示牌行驶约350米即达。
自驾线2：机场高速—花山出口—入花山镇—在花山医院旁边路口左转直入至目的地。

Address: Tieshan Village, Huashan Town, Huadu District
Tel: 020-86788172   Ticket: 100 Yuan
Bus route: Take the No.4 bus to the terminus and then a motor to the destination.
Self-driving route 1: Airport Expressway - Exit of Huadu - Yingbin Avenue - G106 - X284 (a 2-kilometer drive along X284 before turning right) – 350 meters away from the destination by car according to the signpost.
Self-driving route 2: Airport Expressway – Exit of Huashan – Enter in Huashan Town – Turn left on the road beside Huashan Hospital and drive without any turning to the destination.

GUIDE TO HUADU

花都全攻略 花山镇/**HUASHAN TOWN**

👁 福源水库
FUYUAN RESERVOIR

福源水库在花山镇两龙圩北8千米处。太平天国领袖洪秀全出生于当时花县的福源水（今福源水库一带），因而花山又得名"天王故里"。福源水库山清水秀，绿树成荫，每年禾雀开花时节，吸引众多游客慕名前往观赏。

Fuyuan Reservoir is located 8 kilometers away from the north of Linglong Dyke in Huashan Town. Hong Xiuquan, the leader of the Taiping Kingdom of Heaven against the Qing Dynasty, was born in the then Fuyuan Water, Hua County (now the Fuyuan Reservoir area), hence Huashan Town is also known as the "Hometown of the King". With beautiful mountains, clear waters and shade-making trees, Fuyuan Reservoir attracts numerous tourists for sightseeing during the blossom of Dream Bridwood's Mucuna each year.

⭐ 推荐看点
RECOMMENDED VIEWING FOCUS

福源水库周边有外形奇特的禾雀花，又名白花油麻藤、雀儿花。每年3～4月下旬开花，其花形酷似雀鸟，吊挂成串有如禾雀飞舞。花开在藤蔓上，吊挂成串，每串二三十朵不等，串串下垂，有如万鸟栖枝，神形兼备，令人叹为观止。

Dream Bridwood's Mucuna with peculiar shape, grown around Fuyuan Reservoir, is also known as Mucuna birdwoodiana. It blossoms in late March or April in the shape of birds, hung in strings as if java sparrows are dancing in the air. Bloomed in the vines, the flowers are hung in strings with twenty to thirty each to form an amazing picture that all these strings are hung down similar in both externality and internality to thousands of birds perching on the branches.

花山碉楼群
Watchtower Complex in Huashan

　　花山镇是侨乡，能看到诸多华侨建筑，这些建筑大部分的布局、内部装修与陈设以中式为主，建筑造型则以西式为主，形态庄稳，装饰华美，兼具中西建筑风格，为研究近代建筑艺术提供了良好的实例。洛场村以"起鹏楼"为代表的华侨楼群以及平山村、和郁的碉楼群远近闻名，为花山创下了丰富的华侨文化。

As a hometown of overseas Chinese, Huahsan Town has a lot of architectures with overseas Chinese characteristics. The overall layout, internal renovation and furnishings of these architectures are mainly in Chinese style while the architectural style mainly western, which are modest and solemn in shape and resplendent in decorations, combining together Chinese and western styles as well as providing a good example for studying modern architectural arts. The overseas Chinese architectural complex represented by "Qipeng Building" in Luochang Village and watchtower complex in Pingshan Village and Heyu Village are known far and wide, originating rich culture of overseas Chinese for Huashan.

★ 推荐看点
Viewing Focus of Recommendations

　　花都洛场村，村中散落着数十栋外观功能类似开平碉楼的华侨屋，是集防卫、居住和中西建筑艺术于一体的乡土建筑群体。花都碉楼多数是由一家一户独立兴建的，所以身形较之开平碉楼要精巧些。碉楼较为分散，需要慢游品鉴，行走之中，你会发现有很多如彰柏家塾一样的精美建筑散落其中。

Luochang Village in Huadu, with tens of houses of overseas Chinese similar in appearance and functions to Kaiping watchtowers scattering, is a vernacular architectural complex combining defending and dwelling functions with Chinese and western architectural arts. Most of the watchtowers in Huadu are built independently by a single house, so they are more exquisite in shape compared with those in Kaiping. It takes time and patience to appreciate these comparatively scattering watchtowers, during which you will find many delicate architectures like "Zhang Bai Family School" scattering in them.

游览攻略
Travel Guides

地址：花山镇洛场村
自驾线：机场高速—往白云机场、从化、韶关方向—花山出口—路口就是洛场村。
Address: Luochang Village, Huashan Town
Self-driving route: Airport Expressway – To Baiyun Airport, Conghua and Shaoguan – Exit of Huashan – Luochang Village is at the junction.

GUIDE TO HUADU

花都全攻略 花山镇/HUASHAN TOWN

花都好吃

花仙荟
GATHERING OF FAIRIES

名副其实的"花仙"荟萃，用香草世界园区中种植的花草作为原料做菜、煲汤、打火锅，食材新鲜，口感独特，有益身心。

必尝推荐：千日红煲土猪骨汤、香草吊烧鸡、竹香莲藕夹、泰式柠檬蒸鲩鱼、上汤天然野菜、香草火锅、粤式香草美食以及花山特色农家菜。

It's a genuine gathering of "Fairies". Dishes, soup and hotpot are prepared with flowers and plants in gardens within the Vanilla World as raw materials, featuring fresh ingredients, unique taste and good for body and mind.

Recommended delicacies: Stewed Local Pork-bone Soup with Gomphrena Globosa, Deep-fried Crispy Chicken with Vanilla, Fried Lotus Root Pie with Bamboo, Thai Steamed Grass Carp with Lemon, Natural Potherb in Broth, Vanilla Hotpot, Cantonese food made with vanilla and special farm food in Huashan.

地址：铁山村，香草世界景区内
订座电话：020-86788772
Address: Vanilla World, Tieshan Village
Reservation hotline: 020-86788772

Gourmet in Huadu

湖山枇杷园
LOQUAT ORCHARD IN HUSHAN

　　花山镇花城村湖山枇杷园，近年来都会举办枇杷节，让游客入园采摘新鲜枇杷。
　　必尝推荐："饮养源"枇杷酒，农场自供的枇杷鸡和枇杷猪是从小用枇杷喂养生长，肉质鲜美且肥而不腻。

Loquat festivals are held in recent years in Loquat Orchard in Hushan, Huacheng Village, Huashan Town, allowing tourists to pick up fresh loquats in the orchard. Recommended delicacies: "Origin of Nutrition" loquat wine; chicken and pork which are tasty and fat but not too greasy because chicken and pigs provided by the farmyard are fed with loquats all along

地址：机场高速—街北高速—花山出口下高速直走—106国道（梯面方向）—直走3千米左右见从化北兴方向指示牌右转—100米左右见花城村牌坊左转—进入邝宗佑纪念大道，直入1千米见枇杷节指示牌左转—沿指引300米到达。
Address: Airport Expressway – Jiebei Expressway – drive straight after getting off the expressway at exit of Huashan – G106 (to Timian) – drive straight for about 3 kilometers and then turn right when seeing a signpost (to Conghua and Beixing) – drive for about 100 meters and then turn left when seeing the memorial archway of Huacheng Village – to Kuangzongyou Memorial Avenue, drive straight for one kilometer and then turn left when seeing the signpost of Loquat Festival – drive for 300 meters following the guide to arrive at the destination.

龙口饭店
LONGKOU HOTEL

　　龙口饭店位于花都区花山镇，环境幽雅，景色秀丽，居住身心备感惬意。
　　必尝推荐：各式美味农家菜。

Located in Huashan Town, Huadu District, Longkou Hotel boasts a quiet and elegant environment, a beautiful view and brings joy and easiness to guests both mentally and physically. Recommended delicacies: various farmhouse delicacies

地址：花山镇第二工业区前进50米　　订座电话：020-86958963
Address: Go ahead for 50 meters from the Second Industrial Park in Huadu Town
Reservation hotline: 020-86958963

GUIDE TO HUADU
花都全攻略 花山镇 / HUASHAN TOWN

花都好吃

美东粥苑

MEIDONG PORRIDGE RESTAURANT

环境典雅华丽，古色古香，颇有农家山庄风味。顶楼露天设计，古朴的水景园林与回廊让人仿若置身世外桃源。

必尝推荐：炆鱼、炆鸡、粥水鸡、粥水鱼。

With an elegant and gorgeous environment, the restaurant of antique style is much similar to a farmhouse. The top floor adopts an open-air design. Guests seem to be brought to a land of idyllic beauty by the pristine waterscape gardens and winding corridors.
Recommended delicacies: Braised Fish, Braised Chicken, Chicken Porridge Hotpot, Fish Porridge Hotpot.

地址：三东大道铁山河路（美东工业园内）
订座电话：020-86786666
Address: Tieshanhe Road, Sandong Avenue (in Meidong Industrial Park)
Reservation hotline: 020-86786666

Gourmet in Huadu

南门饭店
NANMEN RESTAURANT

南门饭店位于106国道旁平西村南门路口，二十多年来一直坚持使用优质食材，菜式做法传统简单，以农村做法酿造出的家乡醋具有消脂、美容、降压、开胃消食等功效；姜煎蛋、芋头粥、家基面豉蒸花腩、头菜肉饼等本地家乡菜式让人饭量大增。花田旁的露天雅座感受着桂花清香，环境虽简约，却以其优良的出品和平价吸引顾客。

Located at the intersection of Nanmen Road in Pingxi Village beside G106, Nanmen Restaurant insists on high quality ingredients and traditional and simple cookery in terms of dishes for over 20 years. Native vinegar brewed in a way adopted in the country has functions such as fat burning, beauty maintaining, depressurization, appetizing and digestion helping; local home-cooked dishes such as Ginger Omelet, Taro Porridge, Home Style Steamed Streaky Pork with Floured Bean Source and Steamed Mustard and Meat Pie give people good appetite. Customers can enjoy the faint scent of osmanthus sitting in the comfortable seats in the open air beside the flower field. Though with an austere environment, the restaurant attracts customers by its excellent quality and fair price.

地址：花山镇106国道旁平西村南门路口
电话：020-86958985

Address: (at the intersection of) Nanmen Road in Pingxi Village beside G106 in Huashan Town
Tel: 020-86958985

花东镇 HUADONG TOWN
花都全攻略 OVERALL GUIDE FOR TRAVEL IN HUADU

花东镇地处珠三角北部,是广州市的北郊,位于花都区的东部,是广州新国际机场的所在地。花东镇名气大,水口营的探花村、格木林、旗杆夹,莘田二村的八角庙,七星村的大王爷庙(云山宫)、狮前村的世外桃源,国家4A旅游景区九龙湖度假区,花侨的十里水果长廊、飞龙山公园、东南亚文化风情,高溪村、港头村的古建筑,九湖村的农会旧址;四联村的孔雀养殖场,李溪村的石硖龙眼园、流溪河拦河坝,高溪村的蔬菜基地,京塘村的莲藕,莘塘的红蜜杨桃等,尤其是花东的古村落集群、明清民居建筑集群更是藏于深闺中,是花都一笔人文旅游的财富,更是研究古花都生活与民情的一个最好范本。

Sitting in the north of Pearl River Delta, as the northern suburb of Guangzhou and the eastern part of Huadu District, the widely-known Huadong Town is where Guangzhou New International Airport locates. Tanhua Village, erythrophleum fordii forest and Qiganjia in Shuikouying, Octagonal Ancient Temple in Xintian No.2 Village, Great Royal Highness Temple (Yunshan Palace) in Qixing Village, Shiwaitaoyuan (Utopia-like place) in Shiqian Village, the National AAAA Tourist Attraction Dragon Lake Resorts, ten-mile fruits corridor in Huaqiao, Feilongshan Park, Southeast Asian cultural aroma, ancient architectures in Gaoxi Village and Gangtou Village, former site of peasant association in Jiuhu Village, peacock farm in Silian Village, Shixia Longan Garden and barrage of Liuxi River in Lixi Village, vegetable base in Gaoxi Village, lotus root in Jingtang Village, cherries and carambolas in Zitang, and especially the ancient village cluster and residential architectures in the Ming and Qing Dynasties which hide deeply in Huadong, all these are a treasure of cultural tourism for Huadu as well as the best example for studying life in Huadu and conditions of its people.

数字看花东 FIGURES OF HUADONG TOWN

花东镇位于广州市花都区的东部,东经113°17′-113°23′,北纬23°23′-23°31′。

Huadong Town is located in the east of Huadu District, Guangzhou, with east longitude 113°17′-113°23′and north latitude 23°23′-23°31′.

GUIDE TO HUADU

花都全攻略 花东镇/HUADONG TOWN

九龙湖度假区
DRAGON LAKE RESORTS

　　九龙湖度假区是国家AAAA级旅游景区，群山环抱水库，山势迤逦回环，群峰叠起争奇，森林茂密连绵，空气清新怡人，飞禽走兽不绝；水库九弯十曲形似蛟龙游动，所以有"九龙湖"之称。九龙湖碧波荡漾，烟波浩淼，水深而辽阔。湖水清冽，可直接饮用，是广州市备用水源之一。九龙湖水库大坝高达50米，长100多米，水面面积2.4平方千米，水底最深处离水面有50米，相当于十多层楼，是花都区最大的水库。水库隐藏在群山之中，就像一条游龙，总长数千米，坐在游艇上，就好像处于弯曲的江河之中，明明已经到达水的尽头，一转弯，还在水中飘荡，真应了古诗所说"山穷水尽疑无路，柳暗花明又一村"。

　　广州九龙湖高尔夫球场是"2010年广州亚运会高尔夫竞赛及训练场馆"，是目前广州最好的高尔夫球场。球场的设计巧妙地将天然湖泊、山谷地势完全融入整个球场，从而成就湖泊球场、山地球场以及兼具两者特点的新9洞灯光球场。

As a National AAAA Tourist Attraction, Dragon Lake Resorts is surrounded by tortuous mountains whose peaks rise one after another to compete and where the forests are continuous and thick, air is fresh and clean and birds and animals come and go in an endless stream. The lake gets the name because the reservoir winds and twists like a flood dragon snaking along and has a rippling surface as well as vast and deep waters covered with mist. Water in the lake is clear and cool, which can be drunk directly and thus being one of the alternative portable water sources in Guangzhou. The dam on the reservoir is over 50 meters high, over 100 meters long and the water in it covers an area of 2.4 square kilometers with the deepest site 50 meters away from the surface, which equals to the height of a building of more than 10 floors, being the largest reservoir in Huadu District. Dragon Lake Reservoir hides in mountains with a length of several kilometers, just like a swimming dragon. When sitting in the yacht, one will find oneself in a winding river and still drifting on the water once turning even if it seems to be the end of the river, as described in an ancient poem "where hills bend, streams wind and the pathway seems to end, past dark willows and flowers in bloom lies another village."

As the "Golf Competition and Training Stadium of 2010 Guangzhou Asian Games", Guangzhou Dragon Lake Golf Course is the best golf course in Guangzhou. The design of the course tactfully integrates natural lakes and valleys into the whole course to form lake courses, hilly courses and new 9-hole floodlit courses integrating the former two characteristics.

★ 推荐看点
RECOMMENDED VIEWING FOCUS

　　九龙湖公主酒店，步入其中，嫣然就是一座纯美的欧洲小镇，其是2010年广州亚运会指定接待酒店。酒店现有豪华欧洲小镇客房、豪华公主房和皇室房、商务套房、行政套房、总统套房、独栋别墅客房等总计331间。其中1号楼群客房更以其最小70平方米的客房面积、挑高楼层4米、独立开阔的景观阳台等被誉为广州最奢华客房的典范。酒店配备一流的康体娱乐设施，5,000平方米的龙泉水疗馆，引用得天独厚的九龙湖天然山泉水，有中、泰、日、印度、巴厘岛五种风情水疗，让游客享受一个放松身心的旅程；拥有室内外游泳池、健身房、网球场、篮球馆、瑜伽馆等运动健身设施，是休闲健身的绝佳场所。

Entering Dragon Lake Princess Hotel which was the receiving hotel specified by 2010 Guangzhou Asian Games, one will feel like entering into a purely beautiful European town. There are 331 in total guest rooms such as luxury European town guest rooms, luxury princess guest rooms, royal house guest rooms, business suites, executive suites, presidential suites, single-family villa guest rooms, among which the guest rooms in the No.1 Building are hailed as model of the most luxurious guest rooms in Guangzhou with a minimum area of 70square meters, 4-meter elevated space and independent and open landscape balcony. The hotel is equipped with first-class health and fitness recreational facilities, a 5000square meters dragon spring spa pavilion which brings in advantaged natural spring water from the Dragon Lake and spa in five styles such as Chinese, Thai, Japanese, Indian and Bali style to allow guests to experience a relaxed journey, exercise and fitness facilities such as indoor and outdoor swimming pools, gymnasium, tennis court, basketball gym and yoga club, being an ideal place for recreation and body-building.

游览攻略
TRAVEL GUIDES

地址：花东镇九龙湖度假区（靠近新白云国际机场）
电话：020-36909888
公交线：广州白云国际机场接驳巴士。
自驾线1：机场北出口—机场高速北沿线（北兴方向）—九龙湖度假区。
自驾线2：广州机场高速—机场高速北沿线（北兴方向）—九龙湖度假区。
自驾线3：京珠高速—北兴出口—九龙湖度假区。

Address: Dragon Lake Resorts in Huadong Town (near the new Baiyun International Airport)
Tel: 020-36909888
Bus route: Shuttle bus in Guangzhou Baiyun International Airport.
Self driving route 1: North Exit of the airport – north line of Airport Expressway (to Beixing) – Dragon Lake Resorts.
Self-drving route 2: Guangzhou Airport Expressway (to Beixing) – Dragon Lake Resorts.
Self-driving route 3: Beijing-Zhuhai Expressway – Exit of Beixing – Dragon Lake Resorts.

GUIDE TO HUADU
花都全攻略 花东镇/HUADONG TOWN

● 高溪村 GAOXI VILLAGE

高溪村据传是欧阳族人于清嘉庆三年（1798年），从白云区江高镇沙溪村迁来此地建成，距今已有200多年历史。有欧阳宗祠一座，民居40座，建筑共41座。高溪村历史文化保护区以古村落、田园风光为主体，被流溪河灌溉渠划分为南、北两片。聚落的空间系统基本完整、自然景观富于特色。保护区建筑群主要以村面建筑为主。村面建筑保存较好，排列整齐。

完整的古村落风貌、天人合一的生产和生活方式、融入生活与艺术的书院文化，娟秀而发达的水系，有着良好的自然景观环境，其独特的生态环境关系可以概括为四个层次：山体村落水体田地，从而形成了独具特色的田园结构；山体作为村落的背景与天然屏障，村落建筑布置在中心位置，是为整个区域的核心；鱼塘等水体围绕村落布置，水系在高溪村不仅是景观的节点，也是生态系统的关键部分，是生态系统的载体；鱼塘区以外为田地圈层，以水田为主。层次关系明确而又相互渗透，从核心到最外圈层按照对自然生态系统的人为影响由强到弱布置。

It is told that the 200-year-old village was built by the Ouyang family in the third year of Emperor Jiaqing in Qing Dynasty (1798) since they relocated here from Shaxi Village Jianggao Town, Baiyun District. There are 41 buildings including an ancestral temple of the Ouyang family and 40 dwellings. The historical and cultural protection zone in Gaoxi Village focuses on ancient villages and rural scenery and is divided into two areas by the irrigation canal of Liuxi River. The village has a basically complete settlement space system and distinctive natural landscape. Architectural complex in the protection zone are mainly village architectures which are well conserved and orderly arranged.

The village has complete style and features of ancient villages, production and lifestyle which follow the concept of unity of man and nature, culture ancient academies integrated into life and art, graceful and well-developed water system, good natural landscape environment and distinct ecological environment which can be summarized into four levels – mountains, villages, water and farmland – to form a unique rural structure. Mountains are a background and natural barrier for villages which are arranged in the center as the core of the whole area. Water system which is arranged around villages such as fishponds is not only a section of landscape, but also a key part in as well as the carrier of ecological system. Outside the fishpond area are layers of farmland, paddy filed mainly. Relationship among these levels is clear and interpenetrated. The levels from core to the outermost level are deployed from stronger artificial impact on natural ecological system to weak one.

推荐看点
RECOMMENDED VIEWING FOCUS

高溪村中的祠堂、家庙、府第、更楼、民居等形式的建筑，布局严谨，保存较完好。以民居为例，以家塾西侧第一列最为完整，现还能看到"天官赐福"、"骈臻百福"、"千祥云集"等字样，照壁上刻有梅花鹿、麒麟、喜鹊、双狮、牡丹等表示吉祥、喜庆的图案，雕刻生动传神，工艺精湛。

Architectures such as ancestral temples, family temples, mansions, watch towers and dwellings are rigorously laid out and well preserved. Take dwellings for example, the first column in the west of the family school is the most complete with Chinese characters such as "Tianguancifu (a gift of happiness)", "Pianzhenbaifu (advent of various blessings)", "Qianxiangyunji (gathering of auspiciousness)" clearly seen. Patterns representing auspiciousness and jubilancy such as sika deer, Chinese dragons, magpies, double lions and peonies are carved on the gate-facing wall. The carvings are vivid and lifelike with exquisite workmanship.

游览攻略
TRAVEL GUIDES

地址：高溪村南靠白云机场，北接机场高速北延线和山前旅游大道，距白云机场北出口仅0.8千米。

Address: south of Gaoxi Village, near Baiyun Airport, connecting in the north the northern extension of Airport Expressway and Shanqian Travel Avenue, only 0.8 km away from the North Exit of Baiyun Airport.

GUIDE TO HUADU
花都全攻略 花东镇/HUADONG TOWN

👁 港头村
Gangtou Village

　　港头村坐北向南,东南西三面有水环绕,素有"三水朝北,四水归源"之美誉。其地理位置优越,是古时花都的水陆交通要道。尤其是村前的流溪河,是广州与北部地区联系的主要水路,在河边建有货运码头,以前村民的经济来源以货物运输为主,大量的货物从这里进出,生意十分兴旺。

Leaning the north and facing the south, Gangtou Village is surrounded by water in east, west and south, and hailed as "three waters facing the north and waters gathering from far and near". Possessing an advantaged geographic location, it was sea and land thoroughfare in ancient Huadu. Liuxi River in front of the village is the main waterway connecting to northern areas by Guangzhou, by which there is a wharf. In the past, villagers made money mainly by cargo transportation, so a large quantity of goods got in and out from the prosperous wharf.

⭐ 推荐看点
Recommended Viewing Focus

　　港头村是现存少见的典型广府民居布局:坐北向南,村中有十一条古巷道,村前为池塘,村后是小山丘,河流、小溪环抱村子。这种布局就像一把梳子,故称"梳子式布局"。布局整齐的村落、三间两廊式的三合院、村前的池塘、村中的祠堂,还有锅耳式的山墙、灰塑梁脊艺术。村中古建筑以始祖文孙曾公祠为中轴线,向东西两旁延伸,均为明清两代建筑风格,现保存完好的古建筑有60座,其中祠堂、书院、厅堂6座,其余为民居。

With a typical Cantonese dwelling layout which is rare among existing ones, Gangtou Village leans the north and faces the south. There are eleven ancient laneways in the village surrounded by rivers and creeks, of which the front is a pond and the back a hill. This layout is named "comb layout" because it is like a comb in shape: orderly arranged villages, three-section compound with three rooms and two corridors, pond in front of the village, ancestral temple in the middle, handle-like gables and lime-modeling beams and ridges. With the temple of ancestor Zeng Wensun as the central axis, Ancient architectures which adopt architectural styles in the Ming and Qing Dynasties extend to east and west. There are 60 well protected ancient architectures, among which 6 are temples, ancient academies and halls while others are dwellings.

📝 游览攻略
Travel Guides

　　地址:花都大道港头村小学对面村道进入,直入达港头村村委会,即到古民居建筑群。

Address: Huadu Avenue – enter from the village road opposite to Gangtou Village Primary School – straight to the Village Committee, also to the ancient dwelling architectural complex.

👁 探花故里水口营

水口营村有"连科三进士,同榜两贡生"之美称。商家自商衍鎏的父亲商廷焕开始,一门四代从文,连中两位进士。所以有人说,商家是近现代广东"书香门第"的代名词。中国最后一位"探花郎",便是该村的商氏后人商衍鎏。

Shuikouying Village has enjoyed the reputation of "three Jinshi (scholars) in three consecutive imperial examinations and two students studied in the Imperial College on the same list". Starting from Shang Tinghuan, Shang Yanliu's father, four generations of the Shang's family were devoted to literature and two Jinshi got out from the family in succession. As a result, the Shang's family is said to be synonym for "a scholarly family" in modern Guangdong. The last "Tanhua (rank the third in the imperial examinations)" of China, Shang Yanliu, was a descendent of the Shang's family in the village.

★ 推荐看点
RECOMMENDED VIEWING FOCUS

去水口营除了探访我国最后一位探花故居、故里之外,还推荐去探访一群珍贵的"格木林"。村北侧滨水沿岸现存一片30亩的格木林。格木是我国珍稀的二级保护古树,格木林连着一片近100亩的鱼塘及近50亩的荔枝园。

Tourists are recommended to visit not only the hometown of China's Tanhua, but also a group of precious "Erythrophleum fordii forests" in Shuikouying Village. In the north of the village exists a erythrophleum fordii forest covering an area of 30 mu, which is a valuable and rare ancient tree under national secondary protection. The forest is connected with a nearly 100-mu fishpond and a nearly 50-mu litchi orchard.

📝 游览攻略
TRAVEL GUIDES

地址:花东镇花都大道旁。
Address: beside Huadong Town Huadu Avenue

GUIDE TO HUADU
花都全攻略 花东镇 / HUADONG TOWN

广州农科大观
GUANGZHOU AGRICULTURAL SCIENCE INSTITUTE

广州农科大观位于花都区花东镇新白云机场附近，占地400多亩，是全国农业科普教育基地、广东省青少年科技教育基地、省级无公害蔬菜生产基地。其餐厅可为游客提供特色鲜蔬火锅及农家小炒等美食，配套有功能齐全的会议室、舒适干净的客房、多功能卡拉OK等娱乐设施。

With an area of over 400 mu and located near the new Baiyun Airport in Huadong Town, Huadu District, Guangzhou Agricultural Science Institute is a base for national agricultural science education, a scientific and technological education base for adolescent in Guangdong Province and a production base for province-level pollution-free vegetables. Its canteen can provide tourists with delicacies such as special hotpot with fresh vegetables and farmhouse dishes, meeting rooms with complete functions, cozy and clean guest rooms and recreational facilities such as multi-function karaoke.

推荐看点
RECOMMENDED VIEWING FOCUS

农博馆、农趣苑、躬耕园、花卉展示区、无土栽培以及温室种植展示区等，可体验田间采摘蔬菜、鱼池垂钓、自助烧烤等趣味活动。

Agricultural museums, rural fun gardens, experiencing gardens, exhibition areas for flowers, soilless culture and greenhouse planting, etc. Tourists can experience various interesting activities such as picking vegetables in farmland, fishing in the fishpond, self-help barbecue.

游览攻略
TRAVEL GUIDES

地址：花东镇桑梓路高溪村23号（新白云机场北出口）
电话：020-86762648　　门票：20元
公交线：在花都客运站乘坐花都66路（或花都16路），在花东市场站下车，步行1000米至目的地。
自驾线：机场高速—机场北出口—花东大道—桑梓路约2千米处。

Address: No.23 Gaoxi Village, Sangzi Road, Huadong Town (north exit of the new Baiyun Airport)
Tel: 020-86762648　　　　Ticket: 10 Yuan
Bus route: take the Huadu No.66 bus (or the Huadu No.16 bus) at Huadu Passenger Station to Huadong Market and then walk 1000 meters to the destination.
Self-driving route: Airport Expressway – North Exit of Airport – Huadong Avenue – about 2000 meters from Sangzi Road.

狮前村
SHIQIAN VILLAGE

　　狮前村位于花都区东部山前旅游大道以北的山谷中，四面环山，是远离都市喧嚣的"世外桃源"，是临近大都市少有的原生态农村，是个大氧吧。村民全都是客家人，全村人口一千多，现在留在村里的人只有两三百人，保留了客家人的文化风情。目前没有一家工业类的企业，所以没有任何形式的污染，山林、坡地、溪流、田园几乎都处于原生态。

Located in the valley to the north of Shanqian Travel Avenue in the east area of Huadu District, Shiqian Village is surrounded by mountains, being a "Utopia" far away from urban hustle and bustle, a rare original countryside near metropolis and a great oxygen bar. All of the 1,000 more villagers are Hakka people with only 200 to 300 of them stay in the village who retain Hakka cultures and customs. So far, no industrial enterprises have set there. As a result, the village has no pollutions of any type and mountains, forests, slopes, creeks, farms and gardens are nearly in original ecology.

推荐看点
RECOMMENDED VIEWING FOCUS

　　在村子下游处一个叫"鱼惊潭"的地方，有一个高十多米的瀑布。村子里有一个全国最大的中华鲟养殖基地。狮前村周边都是国家的生态公益林，山上的树木维系着九龙潭水库的水源。

　　泡氧吧、登狮山、戏清泉、观美景、赏风光、摘佳果、尝佳肴、购山珍，是狮前村旅游的最佳选择。

At the downstream of the village sits a place called "Yujing (frightened fish) Pond" which collects a waterfall of more than 10 meters high. There is the biggest cultivation base for Chinese sturgeon nationwide. Around Shiqian Village are national ecological public-welfare forests. Trees on the mountain maintain the water source of Jiulong Pond Reservoir.
The best choices for travelling in Shiqian Village includes spending time in an oxygen bar, climbing up Shishan Mountain, playing by clear springs, appreciating beautiful sceneries, picking fresh and good fruits, tasting delicacies and purchasing native delicacies.

游览攻略
TRAVEL GUIDES

地址：花东镇山前旅游大道约9千米处左侧。
Address: about 9km away from Huadong Town Shanqian Travel Avenue, on the left side.

GUIDE TO HUADU
花都全攻略 花东镇/HUADONG TOWN

红色革命摇篮九湖村
CRADLE OF RED REVOLUTION – JIUHU VILLAGE

在九湖村，有一个叫"王氏大宗祠"的地方，是花县第一届农民协会旧址，也是花县农民自卫军总部遗址。称其为红色革命摇篮恰如其分，因为这里曾经是彭湃、阮啸仙、刘尔崧、王福三、王彭等农民运动领导者共同战斗过的地方。

九湖村位于花都大道旁，占地面积约1.5平方千米，全村1100多人。王姓在这里是大族，其宗祠规格自然也很高。这一点，从门口的七级台阶和残存的"马上封侯"的门柱石刻以及三叠三陈的房屋布局可见一斑。据历史记载，十一世命卿祖公，于明朝万历三十三年中举人，癸丑科中进士，出任福建省福清知县，升授湖南长沙县知府，连升刑部主事，转迁礼部正郎，后升朝廷左丞相，其级别大致相当于现在的国务院副总理。

王氏大宗祠为王卿公所立，其建筑古色古香，包括屏风、廊柱、屋檐、斗拱等都很有档次，现仍保存着大量独具岭南风情的灰塑及书法、绘画作品，并残存着一些农民运动时代的照片及简略的文字资料。

In Jiuhu Village, a place called "the Wang's ancestral temple" turns out to be the former site of the first session of peasant association as well as site of the headquarter of peasant self-defense corps in Huaxian County. It is appropriate to call the village the cradle of red revolution because it is the place where leaders of peasant movements such as Peng Pai, Ruan Xiaoxian, Liu Ersong, Wang Fusan and Wand Peng had fighted.

Jiuhu Village is located near Huadu Avenue, covering an area of about 1.5 square kilometers with a population of more than 1100. The Wangs are of great population here, so their ancestral temple has a higher standard, which can be seen from seven-step terraces, remaining stone carvings "Mashangfenghou (to be conferred as a nobleman immediately)" on gate post and the three-layer architectural layout. It is historically recorded that the eleventh generation Progenitor Mingqing passed the imperial provincial civil service examination and became Jinshi (rank the third in imperial examination) on Kuichou subject. He took up the post of the magistrate of Fuqing County, Fujian Province, and was then promoted as magistrate of Changsha County, Hunan Province. After that, he was even promoted as the chief of the Ministry of Penalty, and then the chief of the Ministry of Rites. Finally, he was elevated to the left prime minister of the imperial court, which roughly equals to current vice president of the State Council.

The Wang's ancestral temple was set up by Wang Qinggong, with antique architecture and high-end folding screens, colonnades, eave and brackets. A large number of lime moldings, calligraphy and paintings with Lingnan style as well as remaining photographs and brief documents during times of peasant movements are still kept there.

★ 推荐看点
RECOMMENDED VIEWING FOCUS

花县农民自卫军的前身、自发性的农民革命武装队伍、花县农民战斗的大本营和革命策源地。

The predecessor of Huaxian County Peasant self-defence corps The spontaneous revolutionary armed forces of farmers The farmer revolution base camp and National revolution original place of Huaxian County

游览攻略
TRAVEL GUIDES

地址：花东镇九湖村。
公交线：在花都客运站乘坐花都66路或16路车到花东镇花都大道三凤加油站，步行至九湖村。
自驾线：机场高速—机场北出口—花东镇—花都大道三凤加油站—九湖村。

Address: Jiuhu Village, Huadong Town.
Bus route: Take the Huadu No.66 or No.16 bus at Huadu passenger Station to Sanfeng gas station, Jiuhu Village, Huadu Avenue Huadong Town.
Self-driving route: Airport Expressway–North Exit of Airport–Huadong Town–Sanfeng gas station–Jiuhu Village.

GUIDE TO HUADU
花都全攻略 花东镇 / HUADONG TOWN

莘田二村八角古庙
Octagonal Ancient Temple in Xintian No.2 Village

　　莘田二村八角古庙乃广东省三大新发现之一。该八角古庙始建于明末清初，供奉洪圣爷以镇妖护符。迄今逾600年历史，占地600多平方米。八角古庙建筑独特，结构严谨科学。该庙分为前堂、天井、后堂以及宿舍等。庙堂瓦面有八个角，故称八角古庙。整座庙是廊柱式结构，共32条柱子，石柱4条，其余为木柱。庙的前堂有两幅画廊，后堂一幅画廊。前堂左画廊为"八大仙"，前右画廊是"五鬼运钱程"，后堂画廊为"双龙戏珠"，历史较为久远，最有文物价值。

Octagonal Ancient Temple in Xintian No.2 Village is one of the three new discoveries in Guangdong Province. Built in late Ming Dynasty and early Qing Dynasty, The Saint Hongsheng is consecrated to drive away evils and protect the village. he temple has a history of 600 years and covers an area of more than 600 square meters. With a rigorous and scientific structure, the unique temple is divided into antechamber, courtyard, backyard and dorm. Tiling of the temple has eight angles, thus the name came. With a colonnade structure, the temple has 32 pillars, among which 4 stone ones and 28 wooden ones. There are two corridors with paintings in the antechamber and one in the backyard. The painting in the left corridor of the antechamber is "Eight Great Immortals", that in the right is "Five Ghosts Carrying Wealth and Promotion" and that in the backyard corridor is "Two Dragons Playing a Ball". These paintings have a long history and great value of cultural relics.

推荐看点
RECOMMENDED VIEWING FOCUS

"八大仙"、"五鬼运钱程"、"双龙戏珠"画廊。
"Eight Great Immortals" "Five Ghosts Carrying Wealth and Promotion" "Two Dragons Playing a ball".

游览攻略
TRAVEL GUIDES

地址：花东镇莘田二村
公交线：在花都客运站乘坐66路或16路车到花东北兴转北钟路直入3公里
自驾线：机场高速—机场北出口—花东镇—莘田二村

Address: Xintian No.2 Village, Huadong Town.
Bus route: Take the Huadu No.66 or No.16 bus at Huadu passenger Station to Beixing, Huadong Town, trun to Beizhong, walk along 3km.
Self-driving route: Airport Expressway–North Exit of Airport–Huadong Town–Xintian No.2 Village.

GUIDE TO HUADU
花都全攻略　花东镇 / HUADONG TOWN

花都好吃

 自然美食山庄
NATURAL GOURMET MOUNTAIN VILLA

山庄内设有果园、花卉园、棋牌室、钓鱼台、岭南长廊等休闲、娱乐设施；整个布局具有岭南庭院特色。山庄外还配有养殖场和蔬菜场，其养殖场的走地黑绿鸡远近驰名。

必尝推荐：红烧鸡、精制流溪河鱼系列、白灼黄秋葵、蒜蓉炒薯叶、大头鱼一味四食等。

The villa is equipped with leisure and recreational facilities such as orchards, flower gardens, chess and card rooms, fishing terraces and Lingnan (south of the five Ridges) long corridor, the whole layout featuring courtyards in Lingnan. A livestock farm and vegetable garden are built outside the villa. Local black and green chicken raised in the livestock farm are famous far and near.

Recommended delicacies: braised chicken with soy bean, series of refined fishes from Liuxi River, scalded okra, minced garlic fried with potato leaves, bullhead with four different ways to eat.

 地址：李溪拦坝河堤路上游　　订座电话：13719292308
Address: upstream of Hedi Road at the barrage of Li Stream　　Reservation hotline: 13719292308

 华成餐厅（东南亚风味）
HUACHENG RESTAURANT (SOUTHEAST ASIAN FLAVOR)

华成餐厅（东南亚风味）毗邻广州国际航空港的流溪河畔花都华侨农场内，花都华侨农场因其独特的异国风情，被誉为流溪河畔的"小东南亚"。本餐厅主营东南亚风味，使客人不出国门就可品尝到东南亚美食。

必尝推荐：泰汁焗鱼、越南春卷、印尼咖哩鸡等。

Huacheng Restaurant (Southeast Asian flavor) is located in Huadu Overseas Chinese Farm by Liuxi River near Guangzhou International Airport. The farm is honored as "Little Southeast Asia" by Liuxi River due to its unique exoticness. The restaurant mainly serves dishes of Southeast Asian flavor, allowing customers to have a taste of delicacies from Southeast Asia without going abroad.

Recommended delicacies: baked fish with Thai sauce, Vietnam spring rolls, Indonesia curried chicken, etc.

 地址：华侨农场市场旁（加油站对面）
Address: beside Overseas Chinese Farm Market (opposite to gas station)

Gourmet in Huadu

柳溪河山庄
Liuxi River Mountain Villa

　　山庄内设有果园、花卉园、棋牌室、钓鱼塘、江南长廊等休闲、娱乐设施；整个布局具有江南庭院特色，古香古色，别具一格。山庄外还配有养殖场，其走地鸡、会飞的水鸭都很有名。

　　必尝推荐：烧排骨、煲老水鸭、猪肚煲蛇、宿骨大头鱼、红烧羊肉、虾酱青豆等。

The villa is equipped with leisure and recreational facilities such as orchards, flower gardens, chess and card rooms, fishing ponds and Jiangnan (south of the Yangtze River) long corridor, the whole layout featuring courtyards in Jiangnan which is antique and unique. A livestock farm is built outside the villa, in which local chicken and teals capable of flying are famous far and near.
Recommended delicacies: braised ribs, stewed old teal, stewed port tripe with snake, Aristichthysnobilis, braised mutton with soy sauce, green beans with shrimp sauce, etc.

 地址：山庄在流溪河畔，花东绿道旁边
Address: by Liuxi River and beside Huadong Greenway

陶塘农庄
Taotang Farm Village

　　口碑颇赞的一家农家菜餐厅，做的本土家乡醋尤受推荐，出品质量稳定，性价比较高。

　　必尝推荐：家乡醋、豆豉炆鹅、胡椒鸡。

Winning great public praise, this farm food restaurant sells local home-made vinegar with stable quality and high cost performance, thus especially recommended to customers.
Recommended delicacies: home vinegar, braised goose with bean sauce, pepper chicken

 地址：花东镇桑梓大道（近新机场北端出口）　订座电话：13711081226
Address: Sangzi Avenue, Huadong Town (near the north exit of the new airport)
Reservation hotline: 13711081226

赤坭镇篇
花都全攻略
CHINI TOWN
OVERALL GUIDE FOR TRAVEL IN HUADU

赤坭镇地处花都区西北部，镇内山清水秀、风景宜人、空气清新、自然资源丰富、湖泊水库较多；赤坭是花都区的农业大镇，也是花都区的教育强镇，其"鸦山旭日"、"巴江烟雨"曾为闻名全广州的著名景点。赤坭镇是"全国盆景之乡"，瑞岭村的盆景栽培已有一百多年历史，村中家家户户均有盆景园。

Chini Town is located in the northwest of Huadu District, where there are beautiful mountains, clear waters, pleasant sceneries, fresh air, rich natural resources and a lot of lakes and reservoirs. Chini Town is a big agricultural town of powerful education, whose scenic spots such as "rising sun in Yashan Mountain)" and "misty rain on Bajiang River" was once famous all around Guangzhou. Chini Town is the "Home of National Potted Landscape". Cultivation of potted landscape in Ruiling Village has a history of more than 100 years and each household in the village has a garden for potted landscape.

数字看赤坭
FIGURES OF CHINI TOWN

赤坭镇内土地总面积160.03平方千米，巴江河、省道（S114线）南北贯穿全境，花都山前旅游大道直通佛山市。

The total area of land in Chini Town is 160.03 square kilometers. Bajiang River and S114 run through the whole town south-north. Shanqian Travel Avenue in Huadu directly connects to Foshan City.

GUIDE TO HUADU

花都全攻略 赤坭镇/CHINI TOWN

御盛休闲观光农场
YUSHENG LEISURE AND SIGHTSEEING FARM

御盛观光休闲农场是广州市花都区首个以观光休闲为主题，集跑马、休闲、餐饮、旅游及生态环保教育于一体的休闲娱乐场所。景区以骑马游山为主打项目，引进国内外多种优质品种的马驹，并聘请专业的驯马师加以驯养，现已能满足从休闲到跑马运动等不同层次的多种需求，游客在此可享受沙圈驰骋、骑马游山、垂钓烧烤、星空露营等多种服务。

Yusheng Leisure and Sightseeing Farm is the first leisure and recreation venue in Guangzhou that takes sightseeing and leisure as its theme and combining horse racing, leisure, catering, tourism and education of ecological environment protection. The farm takes touring the mountain by riding a horse as a staple, introducing various high-quality horses and employing professional horse trainers to train these horses so as to satisfy different demands from leisure to horse racing. Tourists can enjoy multiple services such as galloping on sand, touring mountains by riding a horse, fishing, barbecue and camping under the starry sky.

推荐看点
Recommended Viewing Focus

根据欧美专家的研究报告，骑马可以调正姿势体态，训练平衡感、韵律感、柔软度，是一项力与美的最佳运动；有研究数据显示，骑马是最好的健美综合运动，可减肥瘦身，骑马10分钟等于按摩10万次。

According to research reports from Occident experts, horse riding is a sport synthesizing strength with beauty, which can help build correct postures and beautiful physique appearance, train sense of balance and rhythm as well as flexibility. Data shows that horse riding is the best body-building sport for slimming. Riding a horse for ten minutes equals to taking massage 100,000 times.

游览攻略
Travel Guides

地址：赤坭镇锦山村
自驾线：广清高速—赤坭出口—省道114—培正路口—进入锦山村。

Address: Jinshan Village, Chini Town
Self-driving route: Guangzhou-Qingyuan Expressway – Exit of Chini – S114 – intersection of Peozheng Road – enter into Jinshan Village.

GUIDE TO HUADU
花都全攻略　赤坭镇/CHINI TOWN

宝桑园
BAOSHANG GARDEN

宝桑园坐落于花都区山前大道，园区中数十万棵桑树和果树营造出最优美的自然生态环境；春季的桑果成熟期是在3～4月；秋季的桑果成熟期是在11～12月，由于这个季节雨水适当，日晒时间充足，加上宝桑园精心培育，产生了非常难得的反季节的桑果，这个季节的桑果的维生素C含量十分高，是真正的天然绿色食品，被称为"果皇"。

Located on Shanqian Avenue in Huadu District, Baosang Garden has hundreds of thousands of mulberry trees and fruit trees, creating the most gorgeous natural ecological environment. Mulberries become mature in March or April in spring and November or December in autumn. The garden produces extraordinary out-of-season mulberries due to adequate rain, sufficient sunshine and meticulous care taken by staff in the garden. Mulberries in this season contain much Vitamin C to be a genuine natural green food and are hailed as "Emperor of Fruits".

推荐看点
RECOMMENDED VIEWING FOCUS

在宝桑园中,可以在桑林中漫步,随意品尝那红得发紫发黑的挂满枝头的桑果;参观中国5000多年辉煌蚕丝文化历史的丝绸文化长廊,感受举世闻名的丝绸之路;在昆虫天地了解昆虫神奇的奥秘,制作昆虫标本,亲自参与喂养蚕宝宝、缫丝……了解蚕宝宝的一生,体验古诗中"笔落似蚕声"的涵义,还可以在5万平方米的青青草地上体验风靡全世界的水上波波球、草地悠波球的滚筒洗衣机天翻地覆的感觉和太空行走的乐趣。

Tourists can wander in the mulberry forests, taste at ease those dark red mulberries covering branches, visit corridor of silk culture which carries a glorious silk cultural history of more than 5000 years, feel the world-known Silk Road, get to know the magical secrets about insects in Insect World, make insect specimens, participate in feeding silkworms and reeling silk to know about the whole life of a silkworm, experience the meaning of a verse from an ancient poem "the sound of writing is like that made by a silkworm" as well as experience the feeling of reversing the sky and the ground just like roller washing machine by playing water Poepoe ball and grassland Zorb ball on the green grassland of 50000 square meters and the fun of spacewalking.

游览攻略
TRAVEL GUIDES

电话:020-86729815　　门票:48元

花都出发路线:花都市区经建设北路,往芙蓉度假区方向行驶,过杨屋收费站直行至建设北路尽头转左,山前旅游大道(转右往芙蓉度假区),往清远方向直行,过山前旅游大道缠岗收费站后即到。
广州出发路线:广清高速公路,至狮岭山前旅游大道出口出高速。过高速出口后转左,往碧桂园假日半岛方向直行,过山前大道缠岗收费站后即到。
背包族路线:广州市汽车站(流花车站)广州—石角,在缠岗站下车。

Set out from Huadu: drive to the direction of Furong Tourist Resorts via Jianshe North Road in Huadu downtown, turn left before going straight to the end of Jianshe North Road after passing through Yangwu Toll Station, to Shanqian Travel Avenue (turn right to Furong Tourist Resorts), go straight to the direction of Qingyuan, arrive the destination after passing through Changang Toll Station on Shanqian Travel Avenue.
Set out from Guangzhou: drive on Guangzhou-Qingyuan Expressway to Exit of Shanqian Travel Avenue in Shiling and get off the expressway, turn left after getting out from the expressway exit, go straight to the direction of Biguiyuan Holiday Peninsula and arrive the destination after passing through Cangang Toll Station on Shanqian Travel Avenue.
For backpackers: take the Guangzhou – Shijiao line in Guangzhou bus station (Liuhua Station) and get off at Changang station.

GUIDE TO HUADU
花都全攻略　赤坭镇 / CHINI TOWN

瑞岭盆景村
Ruiling Village (famous for potted landscape)

　　瑞岭盆景的历史悠久，至今已有100多年的历史，是作为中国盆景艺术南派的代表。与北方盆景相比，瑞岭盆景以苍劲、自然、飘逸、豪放为艺术特色，在长期发展过程中又形成了自己独特、鲜明的艺术风格，可以用"古、灵、精、怪"四个字做高度的概括。1986年，英女皇伊丽莎白访华时，原广东省省长叶选平把一盆栽种了60年的九里香树桩盆景送给英女皇作为贺礼，这个盆景正是赤坭镇瑞岭村盆景老艺人朱汉祥的作品，瑞岭盆景首次进入英国皇宫，自此声名远播。

With a long history of more than 100 years, the potted landscape in Ruiling represents China's potted landscape art in the south. Compared with those in the north, potted landscape in Ruiling features vigorous, natural, graceful, bold and unconstrained, and gradually forms its own unique and distinct artistic style which can be briefly summarized by four words "ancient, flexible, refined and odd". In 1986, during the Queen Elizabeth's visit to China, Ye Xuanping, the then provincial governor, gave away a kamuning stump potted landscape which had been taken care for 60 years as a gift to the queen. The potted landscape is one of the works by Zhu Hanxiang, an old crafter from Ruiling Village Chini Town. It's the first time for Ruiling potted landscape to enter the Britain's royal palace. Since then, Ruiling potted landscape are well known.

推荐看点
Recommended viewing focus

　　看瑞岭盆景，首推"古树大道"，这是一条集交通、观光、市场三大功能于一体的独具特色的大道。南起清花公路、北接花都山前旅游大道，纵贯瑞岭盆景花卉基地；在古树大道中间和两边绿化带中，引入种植大型古树和树桩盆景，并点缀奇形怪石，是集交易和观光于一体的古树盆景长廊和化石、活化石的历史长廊。

"Gushu Avenue" is the most recommended to appreciate Ruiling potted landscape, which is a unique avenue integrating traffic, sightseeing with market. The south of the avenue origins from Qinghua Road and connects in the north with Shanqian Travel Avenue in Huadu, running through Ruiling potted landscape and flowers base. Large ancient trees and stump potted landscape are planted in the center of Gushu Avenue and greenbelts on both sides which are dotted with bizzare rocks, becoming a corridor of ancient trees and potted landscape as well as a historic corridor of fossils and living fossils that integrate trade with sightseeing.

游览攻略
Travel Guides

自驾线：广清高速—赤坭出口—省道114—培正路口—古树大道。
Self-driving route: Guangzhou-Qingyuan Expressway – Exit of Chini – S114 – intersection of Peizheng Road – Gushu Avenue.

巴江九曲河
BAJIANG JIUQU RIVER

九曲河风景河段为巴江河在赤坭镇域内的上游段，贯穿蓝田、白坭、门口坑等村。曲流、护堤、农田、水塘相映成趣，水流平缓，适合放游竹排。风光旖旎，环境幽静，是理想的观光休闲旅游绿色长廊。

The scenery reach of Jiuqu River is the upstream section within the area of Chini Town, which runs through villages such as Lantian, Baini and Menkoukeng. Winding rivers, dike dam, farmlands and ponds form a delightful contrast. The water is gentle and suitable for drifting by taking a bamboo raft. With lovely scenery and tranquil environment, it is an ideal green corridor for sightseeing, leisure and tourism.

游览攻略
TRAVEL GUIDES

自驾线：机场高速—花都出口—迎宾大道—283县道—广清高速—山前大道出口—381省道—S114省道—行驶约4千米到达九曲河。

Self-driving route: Airport Expressway – Exit of Huadu – Yingbin Avenue – X283 – Guangzhou-Qingyuan Expressway – Exit of Shanqian Avenue – S114 – arrive at Jiuqu River after a 4km's drive.

GUIDE TO HUADU
花都全攻略 赤坭镇 / CHINI TOWN

故乡里
GUXIANGLI

故乡里岭南文化主题公园的修建旨在"重现历史，留住记忆，回归自然，展示文化"，展现清末民初祖辈的生活场景。景区主要分岭南古建筑、祖辈生活展示（含大型祖辈生活物品展与栩栩如生的蜡像情景）、百艺坊、拓展竞技场（分陆上竞技与水上竞技）、农庄、桑基鱼塘与动物园七个区域。

"Guxiangli" Lingnan Cultural Theme Park is built to "reproduce the history, maintain the memory, return to the Nature and display cultures" and to show scenes of life of ancestors living in the late Qing Dynasty and the early Republic of China. The scenic spot is mainly divided into seven areas such as ancient Lingnan architectures, display of life of ancestors (including a large-scale display of living items of ancestors and vivid waxwork scene), workshop of various craftsmanship, outward sports (land sports and aquatic sports), farm village, mulberry fish pond and zoo.

推荐看点
RECOMMENDED VIEWING FOCUS

为了唤醒我们对故里、对父辈们儿时的记忆，祖辈们的生活时光在这里能充分展现，祖祖辈辈日出而作、日落而息，男耕女织，诗礼传家。在故乡里可以看到诸如广绣、烧玻璃、陶瓷、竹编、剪纸和打铁等。具有岭南特色的中华民间绝艺和精彩民俗文艺表演。

"Guxiangli" reminds us of childhood memories about our hometown and the elder generation. The living of ancestors can be fully displayed here. They worked from dawn to dusk, men ploughing and women weaving. Confucian classics and Taoist codes pass from generation to generation. Tourists can appreciate Chinese folk consummate skills and wonderful folk artist performance with Lingnan characteristics such as Guangzhou embroidery, glass burning, ceramics, bamboo weaving, paper-cuts and iron forging.

游览攻略
TRAVEL GUIDES

自驾线：机场高速（太成出口）—迎宾大道—建设北路—山前旅游大道（赤坭方向）。
背包族路线：广州市汽车站（流花车站）广州—石角，在缠岗站下车；或广州市东圃615路线—花都—花都汽车站708路线花都—兴仁，在缠岗站下车。

Self-route route: Airport Expressway (Exit of Taicheng) – Yingbin Avenue – Jianshe North Road – Shanqian Travel Avenue (to Chini).
For backpackers: take the Guangzhou – Shijiao line in Guangzhou Bus Station (Liuhua Station) and get off at Changang Station; or take the Dongpu No.615 bus in Guangzhou – Huadu – take the No.708 bus (Huadu - Xingren) in Huadu Bus Station and get off at Changang Station.

宝桑园食府
Baoshangyuan Restaurant

宝桑园食府中可以品尝几十个品种的特色宝桑宴农家菜，以桑叶、桑果和桑园饲养的兔子、鸡、鸭等为食材，风味独特，不仅食得健康还能吃出"桑"文化。

必尝推荐：鲜沙姜蒸桑叶兔、水晶灵芝鸡、滋补桑兔汤。

One can taste special Baoshang Banquet farm food of tens of varieties which is made with mulberry leaves, mulberries and rabbits, chicken and ducks raised in its mulberry field. With unique flavor, the dishes are not only good for health, but also remind diners of mulberry culture.
Recommended delicacies: Steamed Mulberry Leaf Rabbit with Fresh Galangal, Crystal Braised Chicken with Lucid Ganoderma, Nutritious Mulberry Rabbit Soup.

地址：山前旅游大道缠岗村路段　　订座电话：020-86729815
Address: Changang Village section, Shanqian Travel Avenue
Reservation hotline: 020-86729815

富祥苑
Fuxiangyuan Restaurant

汇聚口味地道的川菜美食及本地特色菜，环境古典优雅，服务热情周到。

必尝推荐：各式经典川菜。

The restaurant collects authentic Sichuan cuisines and local specialties with classical and elegant environment as well as warm and considerate service.
Recommended delicacies: various classic Sichuan cuisines.

地址：山前旅游大道瑞岭村　　订座电话：020-86720088
Address: Lingrui Village, Shanqian Travel Avenue
Hotline for reservation: 020-86720088

沙河农庄
Shahe Farm Village

农家菜的做法很好地紧锁天然食材的原汁原味，让人吃得健康自在，值得一尝！

必尝推荐：沙河粉、各式农家菜。

Farm food is cooked to well keeps the original taste and flavor of natural food materials, good for people's health and bring easiness. It is worth trying it.
Recommended delicacies: rice noodles, various farm food.

地址：山前旅游大道莲珠村路段　　订座电话：13710436578
Address: Lianzhu Village road section, Shanqian Travel Avenue
Reservation hotline: 13710436578

塱头古村
Langtou Ancient Village

炭步镇篇
花都全攻略
TANBU TOWN — OVERALL GUIDE FOR TRAVEL IN HUADU

炭步镇位于广州市花都区西南面,是珠三角经济开发区工业卫星镇,"三高"农业的重要基地,也是著名的侨乡,被誉为"巴江明珠"。

南宋北方居民南迁至此,于河边堆炭外运,渐成圩头,故得镇名。古朴清丽的炭步人杰地灵,炭步人世代勤奋苦读,人文荟萃,"公孙八科甲"、"七子五登科"、"父子两乡贤"的佳话至今仍流传着。全镇文化古迹和自然景观蕴藏丰富,原花县八景之"双峰朝旭"、"巴江烟雨"就在此地,更有汤氏家庙、洪圣古庙等众多人文历史资源。

炭步是名副其实的广东省古村落之乡,省内三大古村落——塱头古村、茶塘古村及藏书院古村形成的古村落集群,气势宏大,古风撩人。最具传统岭南文化特色的古村落与周边优美田园风光及源远流长的巴江文化结合成"古风+田园"的独特景观,价值之高,实属罕见。

Located in the southeast of Huadu District, Guangzhou, Tanbu Town is an industrial satellite town in the economic development zone in the Pearl River Delta, an important base for "Three Highs (high output, high added value and high technological content)" agriculture, as well as a famous hometown for overseas Chinese, being honored as "Pearl of Bajiang River".

Northern residents in the Southern Song Dynasty moved here and piled up charcoal for outward transportation by the river. The charcoal piles gradually formed a wharf, from which the name of the town came. The antique, beautiful and elegant Tanbu Town is the birthplace of a number of outstanding people. People in Tanbu study diligently one generation after another and many of them become talents. Anecdotes such as "Scholar Gongsun getting the first on eight subjects" "five of the seven sons having passed imperial civil examinations" "both the father and son being virtuous" are still widely circulated. The town reserves rich cultural relics and natural landscape. Two of the former eight sights of Huaxian County, "Twin peaks facing the rising sun" and "misty rain on Bajiang River", are located here along with numerous cultural and historical resources such as the Tang's Family Temple and Hongsheng Ancient Temple.

Tanbu Town is a genuine home of ancient villages in Guangdong Province. The ancient village cluster formed by three ancient villages — Langtou Ancient Village, Chatang Ancient Village and Cangshuyuan Ancient Village — are magnificent and antique. Ancient villages with the most characteristics of traditional Lingnan cultures, beautiful rural scenery nearby and long-standing and well-established Bajing River cultures integrate into the unique landscape of "antique plus countryside style". The value of the town is rarely high.

数字看炭步
FIGURES OF TANBU TOWN

炭步镇总面积为113.5平方千米。
Tanbu Town has a total area of 113.5 square kilometers.

GUIDE TO HUADU

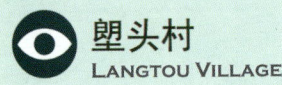 炭步镇/TANBU TOWN

塱头村 Langtou Village

塱头村历史悠久，文化底蕴深厚；全村孕育出较多的有名人物，共有15个进士、10个举人和15个秀才，是名副其实的"进士村"。有着600多年历史的美丽古村，至今仍保留古建筑388座，34间书室、书舍，18条古巷，5座炮楼，较有代表性的有黄氏祖祠、谷诒公书室、乡贤栎坡公祠、友兰公祠、渔隐公祠、留耕公祠等。

作为一座古风优雅的古村落，当地村民至今仍保留着十分淳朴的传统，其民俗展示了"礼仪周全"的古人风雅。比如当地人"做寿"，塱头村定于每年七月二十一日，即始祖忌日第二天为全村60岁及以上老者集体做寿，在祖祠给他们大摆寿宴，并且分猪肉：60岁至69岁的每人1斤，70岁至79岁的每人2斤，如此往上推算增加数量。以往做集体寿宴时只有村里男丁可以参加，如今妇女也可以参加了，但分猪肉已取消，改为封利是（红包），日期也改为了重阳节。再如，当地现在仍尊崇"分鱼"的古法，在年底干塘的时候分鱼给老人。因鱼的大小不一，为防止村民发生分歧，曾有村民建议把鱼连骨带肉煮成鱼酱，并拌入猪油和面豉分给村民。出乎意料，鱼酱鲜香美味。因此，"鱼酱"便成为塱头村祭祖或敬老摆酒席的必备菜式。鱼酱与南乳焖猪肉、豉汁焖牛仔肉合称"塱头三大名菜"。

Langtou Village has a long history and profound cultural deposits. It is a genuine village of Jinshi (ranking the third in imperial examinations), cultivating in total 15 Jinshi and 10 Juren (candidate in provincial imperial examinations) and 15 Xiucai (scholar). With a 600-year history, the beautiful ancient village recently retains 388 ancient architectures, including 34 book rooms, 18 ancient lanes, 5 blockhouses and representative Huang's Ancestral Shrine, Guyi Duke's Book Room, Public Ancestral Temple for Lipo (i.e., Huang Hao, a sage in the county), Public Ancestral Temple for Youlan (i.e., Huang Xueji, the eldest son of Huang Hao) and Public Ancestral Temple for Liugeng.

Living in an elegant and ancient village, the local villagers still keep simple and reserve all the natural traditions and customs which show grace of ancient people -- "comprehensive rites". Take how the local people celebrate their birthday for example: it is scheduled by Langtou Village to celebrate the birthday collectively for the old of 60 years old or elder in the whole village on every July 21st, the second day of death anniversary of the first ancestor, by holding a grand birthday party in the Ancestral Shrine, distributing half a kilogram of pork to age 60 to 69 and one kilogram to age 70 to 79 and so forth. In the past, the collectively held birthday party allowed only male villagers to participate but now female can attend too. Pork distribution has been cancelled and replaced with red packet distribution. Besides, villagers celebrate the birthday on the Double Ninth Festival instead of every July 21st. Another example is that the ancient custom of fish distribution is still popular and fishes are distributed to the old when ponds are dried at the end of each year. Once villages suggested fish sauce (stewed fish with bones) mixed with lard and floured fermented soy bean be distributed to the old to avoid conflicts between villagers due to different size of fishes. To their surprise, the Fish Sauce smells good and tastes delicious. As a result, Fish Sauce becomes a must for ancestor worship or preparing a banquet to respect the old in Langtou Village. Fish Sauce, Braised Streaky Pork with Fermented Red Bean Curd and Braised Veal with Fermented Soy Bean Sauce are called "Three Famous Dishes in Langtou".

推荐看点
RECOMMENDED VIEWING FOCUS

塑头村"三步一书室",可见当时炭步的人文鼎盛之况,如"谷诒公书室",这是塑头村奉直大夫黄谷诒所建的书祠,黄谷诒由于发横财而成为花县的十大财主之一,兴建谷诒公书室和积墨楼,赈灾救民,受封奉直大夫。

进入塑头村之后,推荐以下的旅游路线:百年木棉树—昇平人瑞牌坊—宣重光门楼—渔隐公祠—三馀里古巷—谷诒书室—景徽公祠—履中蹈和门楼—黄氏祖祠—乡贤栎坡公祠—友兰公祠。

The cultural prosperity at that time in Tanbu Town can be seen through "a book room every three steps" in Langtou Village, for example, "Guyi Duke's Book Room" which was built by Huang Guyi, Fengzhi Senior Official (an official position in Yuan, Ming and Qing Dynasty). Huang Guyi struck it rich to become one of the ten rich men in Huaxian County, who built "Guyi Duke's Book Room" and "Jimo Tower (knowledge accumulation)" and devoted himself to relieving disaster and helping others, and was thus granted the Fengzhi Senior Official.
After entering Langtou Village, the following route is recommended: 100-year-old Kapok tree – Shengping Renrui Torri – Sanyuli Ancient Lane – Guyi Book Room –Public Ancestral Temple for Jinghui – Lvzhongdaohe (being moderate) Gate Tower – the Huang's Ancestral Shrine - Public Ancestral Temple for Lipo - Public Ancestral Temple for Youlan.

游览攻略
TRAVEL GUIDES

自驾线:机场高速(太成出口)—迎宾大道—红棉大道—风神大道(炭步方向)。
背包族路线:从广州火车站搭15元班车到炭步镇车站(1小时左右),换乘摩托车。

Self-driving route: Airport Expressway (Exit of Taicheng) – Yingbin Avenue – Fengshen Avenue (to Tanbu Town).
For backpackers: take a shuttle costing 15 Yuan at Guangzhou Railway Station to Tanbu Town Station (about one hour) and then transfer to take a motorcycle.

GUIDE TO HUADU

花都全攻略 炭步镇 / TANBU TOWN

藏书院村
CANGSHUYUAN VILLAGE

　　藏书院位于炭步墟西南6千米处，开基始祖谭嘉靖，于明朝中叶从广州白云区郭塘村分枝迁入发展而成，始名为藏寿庄，建国后行政村名曾改为藏峰村，改革开放后恢复藏书院村名延用至今。村中古村落建筑面积约占全村建筑的三分之一，明清两朝已形成，由庙宇、炮楼、祠堂、书舍、古巷道、古民居构成，整齐统一，保存完好。每列建筑间多有巷里相隔，巷深约195米，现存古巷道11条，大多由石头铺砌，侧砌排水沟，大部分巷里设有水井。巷门楼均嵌红砂岩石额，刻有巷名，其中敦仁里门楼外层嵌红砂岩，里面嵌花岗岩门框，有明清两朝特征。

Located at 6km from the southeast of Tanbu Dyke, Cangshuyuan Village was built by Tan Jiajing who moved from the branch of Guotang Village in Baiyun District, Guangzhou in the middle period of the Ming Dynasty. Its initial name was Cangshou Village, and its administrative name was changed to "Cangfeng Village" after the foundation of PRC. After the reform and opening-up, it regained the name of Cangshuyuan Village and uses it till now. Architectures of ancient villages cover one third of the whole in the village, which formed in the Ming and Qing Dynasties and were composed of temples, blockhouses, ancestral temples, book rooms, ancient lanes and ancient dwellings, orderly and well protected. Each column of architectures are mostly separated by a lane which is about 195 meters deep. There are 11 ancient lanes, most of which are paved with stones and equipped with a drainage ditch on one side. Most of the lanes have wells in them. Gate towers of lanes are inlaid with red sandstone inscribed with names of the lanes, among which the "Dunrenli" Gate Tower is inlaid red sandstone externally and granite doorframe internally with characteristics of the Ming and Qing Dynasties.

推荐看点
RECOMMENDED VIEWING FOCUS

　　到藏书院村观光，您会惊诧于村庄的干净卫生。这是一座有书香气息的古村落，至今仍保持着特别的传统习俗：藏书院每年重阳秋祭，时间不在九月初九，而在八月初六，不是分猪肉而是分鱼，不是只限男丁，而是全村男女老幼都有份。每年新春元宵佳节前一天即正月十四日，全村举行投灯、游灯活动，包括外嫁女及其夫婿、子女、朋友、外族群众均可前来参加，吃大围餐几百席，热闹非常。洪拳舞狮也是藏书院一大特色，至今也有悠久历史；藏书院村非常注重教育，自1996年开始，村里就设了学校专线巴士，每天接送村里所有的孩子到镇上学校去就读。

　　参观路线：洪圣古庙—谭氏祖祠—南炮楼—谭氏宗祠—云溪公祠—云山公祠—桂诗书室—北炮楼。

You will be surprised by the cleanliness and hygiene of the village if sightseeing in Cangshuyuan Village. It is an ancient village full of academic atmosphere and keeping special traditional customs. The village conducts autumn ancestor worship on every sixth day of the eighth lunar month instead of ninth day of the ninth lunar month, distributes fishes instead of pork and welcomes people of all ages and both sexes instead of male only. On January 14 in the lunar calendar, the eve of the Lantern Festival, activities such as casting lanterns and lantern parade are held in the village, participated by all villagers including married daughters and their husbands, children, friends, people from outside. They prepare large-scale banquet with hundreds of seats, bustling with noise and excitement. Hung Fist and lion dance is another feature with a long history. Cangshuyuan Village emphasizes on education. Since 1996, the village set up "Special Bus for School" which sends all children to and picks up them from the school in the town.
Sightseeing route: Hongsheng Ancient Temple – The Tan's Ancestral Shrine – Southern Blockhouse – The Tan's Ancestral Temple – Public Ancestral Temple for Yunxi – Public Ancestral Temple for Yunshan – Guishi Book Room – Northern Blockhouse.

 游览攻略
TRAVEL GUIDES

背包族路线： 从广州火车站搭15元班车到炭步镇车站（1小时左右），转乘摩托车。
For backpackers: take a shuttle costing 15 Yuan in Guangzhou Railway Station to Tanbu Town Station (about one hour) and then transfer to take a motorcycle.

GUIDE TO HUADU
花都全攻略 炭步镇 / TANBU TOWN

茶塘古村
CHATANG (TEA POND) ANCIENT VILLAGE

　　茶塘古村位于炭步镇西南禅炭公路西侧，村民多姓汤，汤姓于宋代从南海迁至此，立村约700年，汤字水旁，茶亦为水，塘能容之，故名"茶塘"。村中古建筑现存较为完整的明清建筑约120座，其中庙宇、祠堂、书院、书室共有20多座，村面建筑以庙宇、宗祠及书舍为主，其中洪圣古庙、明峰汤公祠、友峰汤公祠、万成汤公祠等保存较完整。

Chatang Village is located at the west side of Chantan Road in the southeast of Tanbu Town. Most of the villagers have a family name of Tang, who moved here from the South China Sea in the Song Dynasty. So the village has a history of about 700 years. The left side of the Chinese character Tang means water and tea is water too, which can be held by pond. That is how the village name comes from. There are about 120 architectures built in the Ming and Qing Dynasties that are comparatively complete among ancient architectures in the village, including 20 in total temples, ancestral temples, ancient academies and book rooms. Village architectures are mainly temples, ancestral temples and book rooms, among which Hongsheng Ancient Temple, Public Ancestral Temple for Tang Mingfeng and Public Ancestral Temple for Tang Wancheng are relatively completely kept.

★ 推荐看点
RECOMMENDED VIEWING FOCUS

茶塘的洪圣古庙是花都规模最大的庙宇，建筑艺术、文化内涵各方面一流。庙里面所有的木材都是坤甸木，门前立有两条龙柱。古村中最好玩的看点是"华尔街"——财主佬巷足征里，足征里巷是有名的财主佬巷，屋宇建筑十分讲究，统一为三间两廊，一式青砖，像广州西关大院那样，巷门加"抵制严嵩门"，大厅"趟栊"。村落中一年最热闹的是每年正月十六日那天早上8时的"抢炮会"，各房亲醒狮队陆续进场，各自带上雄鸡、四时果品，先到洪圣古庙拜祭，然后在庙前空地献艺，表演武术，各式器械、拳种、套路令人目不暇接。

Hongsheng Ancient Temple in Chatang is the largest temple in Huadu with the first-class architectural art and cultural connotations. All the woods in the temple are from Pontianak. In front of the temple gate, there stand two dragon pillars. The most interesting focus in the ancient village is "Wall Street" – Zuzhengli, the famous "Lane of the Rich", where house architectures are exquisite. Any of these houses have three rooms and two corridors, paved with black bricks just like West Gate Courtyard in Guangzhou, the lane installed with "door resisting Yan Song" and the hall with "Tanglong Door" (an ancient security door). The most hilarious moment in the village is the "Meeting of Snatching by Crackers" at eight o' clock on the morning of every January 16th in the lunar calendar. Lion dance teams from each family arrive in succession bringing cocks and fruits in four seasons. They worship ancestors in Hongsheng Ancient Temple first and then give performances on the open space in front of the temple, such as martial arts, with various instruments, types of fists and series of skills and tricks making spectators dizzy.

游览攻略
TRAVEL GUIDES

背包族路线：从广州火车站搭15元班车到炭步镇车站（1小时左右），转乘摩托车。

For backpackers: take a shuttle costing 15 Yuan in Guangzhou Railway Station to Tanbu Town Station (about one hour) and then transfer to take a motorcycle.

GUIDE TO HUADU

花都全攻略 炭步镇/TANBU TOWN

花都好吃

家乡渔村
SPECIALTIES OF HOME FISHING VILLAGE IN TANBU

驰名姜葱鸡、南瓜扣芋头
（Famous Ginger and Scallion Mixed Chicken）

土猪肉炆芋头、支竹羊腩煲
（Braised Local Pork with Taro and Dry Tofu Lamb Clay Pot）

 地址：花都区炭步镇鸭湖村　　电话：020-86736252
Address: Yahu Village, Tanbu Town, Huadu District　　Hotline: 020-86736252

美丽农庄
Specialties of Beautiful Farm Village in Tanbu

芋丝蒸腊味、芋头炆水鸭
（ Steamed Sliced Taro with Cured Meat and Braised Taro with Teal ）

牛仔肉、芋头炆五花肉
（ Veal and Braised Taro with Streaky Pork ）

鱼 酱、南乳炆猪肉
（ Fish Sauce and Braised Pork with Fermented Red Bean Curd ）

 地址：花都区炭步镇塱头古村   电话：13822272433（何生）
Address:Langtou Village,Tanbu Town,Huadu District   Hotline: 13822272433 Mr.He

GUIDE TO HUADU

花都全攻略 炭步镇 / TANBU TOWN

高山小溪
SPECIALTIES OF MOUNTAIN AND CREEK IN TANBU

清蒸山水豆腐、清蒸水库鱼
Steamed Tofu in Sauce and Steamed Fish (from the reservoir)

肉丸酿节瓜、秘制叉烧
Stuffed Meatballs with Zucchini

地址：花都区风神大道（四角围加油站附近）
电话：020-86709668
Address: Fengshen Avenue, Huadu District (besides Sijiaowei gas station)
Hotline: 020-86709668

炭步槟榔芋标准化示范基地
Tanbu Betel Nut Taro Standardization Demonstration Base

以文冈村为主要示范区，以种植槟榔芋、粉葛、葱蒜为主。其中种植槟榔芋已有500多年历史，种植面积约400公顷，制定了统一的质量标准和生产技术规范，并建立了芋头标准化示范区、苗种基地、研发区、加工区、展示区、销售区、接待区等。现时炭步槟榔芋经过多年的精耕细作、科学管理，其品质平均优良。可让游客体验种植、收获，进行科普展览、生态廊道游走等。

The base is a vegetable village which takes Wengang Village as the main demonstration zone and mainly plants betel nut taro, arrowroot, scallion and ginger. It has a history of more than 500 years in planting betel nut taro. The base has an about 400-hectare planting area, has formulated uniform quality standard and technical specifications for production and established standardized demonstration area for taro, seed base, research and development area, exhibition area, marketing area and reception area. The present Tanbu taro has a averagely high quality under intensive cultivation and scientific management for many years. Tourists are allowed to experience planting, harvesting and exhibitions for science popularization, and to wander through the ecological corridors.

GUIDE TO HUADU
花都全攻略 炭步镇/TANBU TOWN

 霸王花种植示范基地
PITAYA FLOWERS CULTIVATION DEMONSTRATION BASE

骆村霸王花种植示范基地建在一个岛屿上，四面环水，景观极佳，配套有科普长廊一个，摘果区30亩，钓鱼区100亩，菜地60亩，可出租给城市居民种菜，让城市居民享受都市农业乐趣。

The base in Luocun Village is built in an island surrounded by waters. With beautiful sceneries, the base has a corridor for science popularization, a 30-mu fruit picking area, a 100-mu fishing area and a 60-mu vegetable field which can be rent to urban residents to plant vegetables and enjoy the fun of urban agriculture.

 游览攻略
TRAVEL GUIDES

自驾线1：机场高速（太成出口）—迎宾大道—红棉大道—风神大道（炭步方向）。
自驾线2：北二环—西二环—禅炭路炭步方向—兴华路。
背包族线路：广州市汽车站搭14元班车到炭步镇车站（1小时左右），换乘摩托车。

Travel guides: Self-driving route 1: Airport Expressway (Exit of Taicheng) –Yingbin Avenue–Hongmian Avenue–Fengshen Avenue (to Tanbu).
Self-driving route 2: North Second Ring–West Second Ring–to Shentan Road and Tanbu–Xinghua Road.
For backpackers: take a shuttle costing 14 Yuan in Guangzhou Bus Station to Tanbu Town Station (about one hour) and then transfer to take a motorcycle.

 火龙果园
DRAGON FRUIT ORCHARD

炭步镇大涡村的火龙果农场，是广州市区内最大的火龙果生产基地。火龙果也称长寿果，其周边有四季采摘园，结合人们喜好开展采摘体验区和浪漫拍摄区，深得游人喜爱。火龙果花香虽不浓，却飘逸，类似昙花，只在6月到10月的夜里开花，凌晨则会凋谢，是名副其实的"夜仙子"。

Dragon Fruit Farm in Dawo Village, Tanbu Town is the biggest dragon fruit production base in Guangzhou. Dragon fruit is also known as longevity fruit. A four-season picking garden is around the base, which provides picking experience area and romantic shooting area to satisfy different demands of tourists and is so popular with the touists. The flowers of dragon fruit do not have a thick fragrance, but they are elegant, blooming at nights from June to October and withering at dawns, which are similar to epiphyllum and the genuine "night fairies".

游览攻略
TRAVEL GUIDES

自驾线1：机场高速（太成出口）—迎宾大道—红棉大道—风神大道（炭步方向）。
自驾线2：北二环—西二环—禅炭路炭步方向—兴华路。
背包族线路：广州市汽车站搭14元班车到炭步镇车站（1小时左右），换乘摩托车。

Travel guides: Self-driving route 1: Airport Expressway (Exit of Taicheng) – Yingbin Avenue – Hongmian Avenue – Fengshen Avenue (to Tanbu).
Self-driving route 2: North Second Ring – West Second Ring – to Shentan Road and Tanbu – Xinghua Road
Backpackers: take a shuttle costing 14 Yuan in Guangzhou Bus Station to Tanbu Town Station (about one hour) and transfer to take a motorcycle.

陆仕蝴蝶兰生产基地
LUSHI BUTTERFLY ORCHID PRODUCTION BASE

炭步镇以陆仕蝴蝶兰生产基地为核心，大力推广花卉旅游。群芳争艳，碧野成趣，谁也抵挡不住它的视觉盛宴。春色撩人之时或是傲霜怒放之际的花卉给人无限美感，群花盛放之地更是踏青郊游、休闲观光的绝佳去处，特别是摄影爱好者更是四季追逐她的魅力。

Tanbu Town focuses on Lushi Butterfly Orchid Production Base and promotes flower tourism. Various flowers compete for attention and green grassland form a joyful picture, no one being able to resist this visual feast. Flowers blooming when spring provokes people's interest or by withstanding the frost release endless sense of beauty. The place where flowers blossom is the best for spring outing, leisure and sightseeing, especially to amateur photographers who chase after them all the year round.

游览攻略
TRAVEL GUIDES

自驾线1：机场高速（太成出口）—迎宾大道—红棉大道—风神大道（炭步方向）。
自驾线2：北二环—西二环—禅炭路炭步方向—炭步工业园。
背包族线路：广州市汽车站搭14元班车到炭步镇车站（1小时左右），换乘摩托车。

Travel guides: Self-driving route 1: Airport Expressway (Exit of Taicheng) – Yingbin Avenue – Hongmian Avenue – Fengshen Avenue (to Tanbu).
Self-driving route 2: North Second Ring – West Second Ring – to Shentan Road and Tanbu – Tanbu Industrial Park.
For backpackers: take a shuttle costing 14 Yuan in Guangzhou Bus Station to Tanbu Town Station (about one hour) and then transfer to take a motorcycle.

狮岭镇篇
SHILING TOWN
花都全攻略
OVERALL GUIDE FOR TRAVEL IN HUADU

"狮岭"可以说是花都的一张"皮",其皮革、皮具产业是最有特色的经济、支柱产业、富民产业。从自创品牌到吸引外国知名品牌的落户,成为狮岭自豪的事情,是名副其实的"中国皮具之都"。清乾隆年间(1736—1795年)建圩。因附近有一狮形山岭,故名狮岭圩,镇因圩名。狮岭镇有"广东省盘古文化之乡"称号,也是珠江三角洲地区唯一一个以盘古文化命名的乡镇。

"Shiling" could be praised as the "leather" of Huadu, for the leather and leatherware industry is the most unique economic and pillar industry benefiting local people. Shilling is proud of the original innovation of its own brand and attracting foreign famous brands to settle in Shiling Town, which is worthy of the name of "Capital of Leatherware in China". The dyke was established during the region of Emperor Qianlong (1736-1795) of Qing Dynasty and the town got its name from the mountain ridge in lion shape as "Shiling Xu". It is also reputed the title of "Township of Culture of Pan Gu in Guangdong Province", as well as the only township named as culture of Pan Gu in the Pearl River Delta.

数字看狮岭
FIGURES OF SHILING TOWN

花都区狮岭镇是一个位于广州市北部的美丽城镇,目前镇域总面积160平方千米。

adu District is a beautiful town in the north of Guangzhou, with a total area of 160 square meterssquare kilometers.

GUIDE TO HUADU
花都全攻略 狮岭镇/SHILING TOWN

盘古王庙
TEMPLE OF KING OF PAN GU

每逢农历八月十二盘古王诞这一天，数以万计的村民从四面八方赶至盘古王庙，用各种民间艺术表演争相展示盘古氏开天辟地的英雄风貌。史料记载，"盘古文化"在花都生根已有1500年的历史，迁移到狮岭也有近300年历史。整个南海郡管辖的珠三角地区，只有花都有盘古王山、盘古峒，有关于盘古神坛的史料记载，并在民间广为流传着《盘古王伏龙降狮》等神话传说，1500年前有"南海中盘古国"的南海郡，专家认定南海中盘古国的遗址就在狮岭。

Whenever on the twelfth day of the eighth lunar month each year, the date of birth of Pan Gu King, tens of thousands of villagers rush here from far and wide, with various kinds of folk art performances, to compete to show the hero style of Pan Gu during the creation of the heaven and earth. According to the historical materials, "Culture of Pan Gu" had been in Huadu for 1,500 years and it transferred to Shiling nearly 300 years ago. In the whole Pearl River Delta Region under the jurisdiction of Nanhai County, Mountain of King of Pan Gu and Cave of Pan Gu can only be found in Huadu as well as historical records of Pan Gu Altar. Besides, myths and legends like King of Pan Gu Beat the Dragon and Lion was widely circulated. It is said that Nanhai County was the former "Pan Gu Country in Nanhai" 1,500 years ago, thus, the experts are deeply convinced that the ruin of Pan Gu Country in Nanhai should be in Shiling Town.

推荐看点
Recommended viewing focus

盘古王庙历史悠久，香火鼎盛。清嘉庆十四年（1809年）这里建成的盘古王庙位于盘古王山的山脚，现存的神坛则是清光绪二十七年（1901年）重建的，是广东著名的文化艺术古迹之一。

推荐游览景点：半山亭、盘古卧石、龙口泉、石坪、盘古烟霞等。

Temple of King of Pan Gu, built in the 14th year of Emperor Jiaqing in Qing Dynasty (year 1809) at the foot of Mountain of Pan Gu King, not only has a long history, but also is so popular with visitors. And the existing Altar was rebuilt in the 27th year (year 1901) of Guangxu in Qing Dynasty, is one of the well-known cultural and artistic relics in Guangdong Province.
Landscapes of recommendations: Mid-hill Pavilion, Crouching Stone of Pan Gu, Longkou Spring, Stone Ground, "Haze of Pan Gu", etc.

 ## 游览攻略
Travel Guides

地址：花都狮岭镇振兴村的盘古王山麓
电话：020-86846134　　门票：10元
公交线：花都汽车站—狮岭（每15分钟一班）。

Address: Piedmont of Pan Gu King, Zhenxing Village, Shiling Town, Huadu
Telephone: 020-86846134　　Entrance ticket: 10 Yuan/person
Bus route: Huadu Bus Station- Shiling (15 minutes/shuttle).

GUIDE TO HUADU

花都全攻略 狮岭镇/SHILING TOWN

芙蓉度假区
FURONG TOURIST RESORTS

芙蓉度假区是一个以山、林、湖、泉景观取胜，具有优美的自然景色和神话传说的，人文古迹众多的旅游胜地。芙蓉嶂亦名芙蓉山，海拔360米，重峦叠嶂，方圆数十里，山上石头均表面烟墨色呈芙蓉花形状，故称芙蓉山。

到芙蓉度假区还有可能观赏到"桃花水母"。桃花水母是名副其实的活化石，具有极高的研究价值和观赏价值。花都区成立芙蓉度假区白沙田桃花水母及其生态县级自然保护区，这是广东省第一个桃花水母及其生态县级自然保护区。后来桃花水母又在花山镇磨刀坑水库神秘现身。

Furong Tourist Resorts, famous for landscape of mountain, forest, lake and spring, together with beautify natural scenery, myths and legends, is a tourism attraction with tremendous historical and cultural spots. Furong Peak, also named as Furong Mountain, is 360 meters high, with peaks rising one after another for tens of miles. It is named as Furong Mountain for lotus shape in ink color on the stone surface on the mountain.

Freshwater Jellyfish" is also one of scenery in Furong Tourist Resorts, which is worthy of the name of "living fossil" with extremely high research value and ornamental value. Huadu District established the Baishatian Freshwater Jellyfish and County-level Ecological Natural Reserve in Furong Tourist Resorts, the first freshwater jellyfish and county-level ecological natural reserve in Guangdong Province. Afterwards, the mysterious freshwater jellyfish is discovered in the Modangkeng Reservoir, Huashan Town.

推荐看点
RECOMMENDED VIEWING FOCUS

　　芙蓉度假区是花都的大名片，山上名胜古迹荟萃，林木郁郁葱葱，山南有花都古八景之首的西山瀑布，落差90米。瀑布下面是能万人同时游泳的人工泳池，占地3.3万平方米。池旁是芙蓉嶂水库。芙蓉度假区中的福船岗是北回归线经过之处，树木覆盖率89%，空气清新宜人，有鲤鱼岗、狮山、象山等景点及烧烤区、垂钓区、温泉区、果园等。

Furong Tourist Resorts is called as visiting card of Huadu for places of historic interest and scenic beauty gathering together in the mountain with a wild profusion of vegetation. In addition, Western Hill Fall with 90 meters drop, top of Eight Ancient Scenery of Huadu, is located at south of the mountain. There is an artificial pool covering 33000 square meters under the fall, which is available for ten thousand swimmers, beside the Furong Peak Reservoir. Fuchuan Hillock in the Furong Tourist Resorts is passed by the Tropic of Cancer, with 89% of fraction of forest coverage and clean and pleasant air, in which Carp Hillock, Lion Mountain, Elephant Mountain and other scenic spots is located, together with Barbecue Area, Fishing Area, Spring Area and Orchards etc.

游览攻略
TRAVEL GUIDES

地址：狮岭镇瑞边村芙蓉度假区　　　电话：020-86853355
Address: Furong Tourist Resorts, Ruibian Village, Shiling Town　　Telephone: 020-86853355

GUIDE TO HUADU
花都全攻略 狮岭镇 / SHILING TOWN

狮岭（国际）皮革皮具城
SHILING (INTERNATIONAL) LEATHER AND LEATHERWARE CITY

到狮岭，皮革皮具是永远绕不过去的主题，现在不仅有皮革皮具商贸，更有皮具的文化内涵。狮岭现拥有8个皮革皮具专业市场，具体包括狮岭（国际）皮革皮具城、狮岭（全球）皮革五金龙头市场、宝峰皮革材料交易城、喜龙国际皮革五金采购中心、狮岭国际五金城、纳海皮具饰博园、喜伴城市广场、信宝国际品牌皮具展贸中心等。

You cannot visit Shiling without seeing leather and leatherware, which boasts not only commerce and trade of "leather and latherware", but also cultural connation of leatherware. Currently, there are eight specialized leather and leatherware markets, including Shiling (International) Leather and Leatherware City, Shiling (Global) Leather and Hardware Leading Market, Baofeng Leather Material Business City, Xilong International Leather and Hardware Purchasing Center, Shiling International Hardware City, Nahai Leatherware Upholstery Expo, Xiban Urban Plaza and Xinbao International Brand Leatherware Exhibition and Trade Center etc.

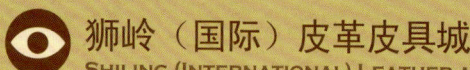

★ 推荐看点
RECOMMENDED VIEWING FOCUS

　　中国皮具之都图书馆是一个汇集了当今皮具产业专业书籍、最时尚的皮具杂志、各种专业影像资料的专业图书馆，它的成立为狮岭皮具设计精英们提供了最前沿的国际潮流信息，也提供了一个资料查询、寻找灵感、学习提升的场所。通过合理运营，目前，中国皮具之都图书馆已有专业藏书近5000册，专业杂志10000册，各类藏书近万册，专业影像碟数百张，极大扩展了狮岭企业家和设计师的国际视野。当地还有狮岭皮具产业研究中心。

China Library of Capital of Leatherware is a specialized library integrating professional books of current leatherware industry, the most fashionable leather magazines and various professional image data, which offers the most cutting-edge international fashionable information for leatherware design elites in Shiling, as well as a site for data inquiry, inspiration pursuit and study promotion. By virtue of rational operation, currently, the library boasts approximately 5000 volumes of professional books, 10000 volumes of professional magazines, almost 10000 volumes of books in various types and hundreds of professional videos, vastly expanding international perspective of entrepreneurs and designers in Shiling. Shiling Letherware Industry Research Center is also located there.

游览攻略
TRAVEL GUIDES

地址：狮岭镇狮岭大道西与宝峰南路交界处
公交线：乘坐花都82路或3路、62路、、81路、80路，在皮革城站下车。
自驾线：机场高速—花都出口—迎宾大道—三东大道西—717乡道—合和路到终点—贵丽路—狮岭皮革皮具城。

Address: junction between Shiling Avenue West, Shiling Town and Baofeng South Road
Bus route: take Huadu No. 82, No.3, No. 62, No.81 or No.80 Bus to Leather City Station.
Self-driving route: Airport Expressway- Huadu Exit- Meeting Avenue- Sandong Avenue West- 717 Village Road- end of Hehe Road- Guili Road- Shiling Leather and Leatherware City.

GUIDE TO HUADU
花都全攻略 狮岭镇/SHILING TOWN

"阳光6号"——中国皮具产业文化创意园
"SUNSHINE NO.6"-CHINA LEATHERWARE INDUSTRIAL AND CULTURAL INNOVATION PARK

"阳光6号"是中国首家皮具产业文化创意主题公园。仅从浅棕与深棕套嵌的建筑外观,以及散落在园区各个角落的皮具箱包主题雕塑,您就能够感受到浓浓的皮具文化气息。园区分为几大功能区,其中包括时尚前卫的中国皮具箱包时尚发布中心,融合了新型多媒体博物馆元素的国内外品牌体验馆,软硬件都一流的皮具箱包观光工厂及培训基地,游客DIY体验区,别致的箱包设计师Loft工作室,具有行业特色的人才市场,皮具箱包创意集市等。

"阳光6号"就是一座以"皮具箱包"为概念、充满活力的博物馆。不管您是大牌设计师、时尚潮人、小资、"小清新",还是皮具箱包行业的行家,都能够在"阳光6号"体验到独一无二的皮具文化。同时,这里还是家庭游乐日的好去处!

"Sunshine No.6" is the first leatherware industrial and innovation theme park. You will experience the rich cultural atomosphere of leatherware from the light brown and dark brown architectural appearanceas well as theme sculptures of leather bags scattering in all corners of the park. The park is divided into several function areas including fashionable and cutting-edge China Leather Bag Fashion Distribute Center, domestic and foreign brand experience pavilion integrating new multi-media museum elements, leather bag sightseeing factory and training base with first-class software and hardware, Visitor DIY Experience Area, unique Loft Studio of bag designer, talent market with industry characteristics, and lather bag innovative market etc.

"Sunshine No. 6" is a vibrant museum with concept of "leather bag". Everyone will experience unique leather culture here, including famous designers, fashionable trendsetters, petty bourgeoisie, people loving fresh style and industry experts of leather bag. Meanwhile, it is a good place for family amusement!

推荐看点
RECOMMENDED VIEWING FOCUS

　　低价购买国内外箱包品牌,高级定制"心水"包包,在创意集市和概念店淘"创意"、淘"设计",亲身体验皮具箱包的生产过程,观赏各种走秀表演,通过新型多媒体技术增长国内外皮具箱包知识,参观设计Loft工作室及露天花园。最后,在阳光咖啡馆品尝美食,欣赏着缘分天河音乐喷泉,享受"阳光6号"给您带来的阳光心情。

Buy domestic and international brand bag at low price, purchase "favorite" advanced customized bags, select "innovation" and "design" in creative market and conceptive store, experience production process of leather bag, appreciate variety of catwalk shows and performances, broaden knowledge in domestic and foreign leather bags through new multimedia technology, visit Design Loft Studio and Outdoor Garden. Finally, enjoy feast in Sunshine Coffee Shop, Yuanfen Tianhe Musical Fountain, and sunny-mood of "Sunshine No. 6".

 游览攻略
TRAVEL GUIDES

地址: 花都区狮岭镇阳光路6号　　电话: 020-86990699
公交线1: 乘车至花都客运站后转乘3、62、80、81、82路公交到文化园站下车,下车后右走100米再过对面即到。
公交线2: 乘坐地铁3号线到人和站B出口转乘701A皮革城下车,改乘摩托车5元或乘坐花91路到文化园站下车即可到达。
公交线3: 乘坐地铁2号线到三元里站A1出口,出站后往右走大概800米到中港皮具城楼下坐"桂花岗—狮岭"直达大巴,到狮岭振兴路十字路口下,过马路往右走约300米即到。
自驾线1: 从广州出发,广清高速海布出口下,下高速后平步大道中调头第一个红绿灯左转进入狮岭大道。约行驶4.2千米后右转进入阳光路,大概200米即可到达。
自驾线1: 从新华出发,沿建设北路往北方向行驶,左转入平步大道中,约2千米后稍向右转进入狮岭大道。约行驶4.2千米后右转进入阳光路,大概200米即可到达。

Address: No.6 Yangguang Road, Shiling Town, Huadu District
Telephone: 020-86990699
Bus route:
Go to Huadu Passenger Station by bus and transfer to No. 3, No.62, No.80, No.81 or No.82 Bus to Cultural Park Station, walk for 100 meters in the right, and across the road.
Go to Exit B of Renhe Station of No.3 metro and transfer to 701A Leather City, and then go to Cultural Park Station by motorcycle (5 Yuan) or taking No.91 Bus.
Go to Exit A1 of Sanyuanli Station by taking NO.2 metro, walk in the right for approximately 800 meters to take "Guihua Hillock- Shiling" Non-stop Coach under Zhonggang Leatherware City to the crossroad of Zhenxing Road, Shiling, across the road and walk in the right for about 300 meters.
Self-driving route 1:Starting from Guangzhou: drive off from Haibu Exit of Guangzhou-Qingyuan Highway, turn around in the Pingbu Avenue, turn left from the first traffic light for approximately 4.2 kilometers and turn right to Yuangguang Road for about 200 meters.
Self-driving route 2:Starting from Xinhua: drive to the north along Jianshe North Road, turn left to Pingbu Avenue for approximately 2 kilometers, turn right slightly to Shiling Avenue for approximately 4.2 kilometers, and turn right to Yuangguang Road for about 200 meters.

GUIDE TO HUADU

花都全攻略　狮岭镇 / SHILING TOWN

👁 广州花卉之都
CAPITAL OF FLOWERS IN GUANGZHOU

广州花卉之都位于狮岭镇，以打造"广州花卉之都"品牌，建设中国南方花卉苗木交易中心为目标，以开展信息交流、促进产业发展为重点，努力办成集园林绿化花卉苗木展示交易、科技信息交流、文化旅游、招商引资与农业生态旅游观光多功能于一体的综合性大型花卉市场。首期2000亩主要功能区分为：花卉主题广场、苗木交易市场、温室花卉区、花卉批发市场、家庭园艺中心、园艺用品交易中心、综合服务中心、花卉资材市场、花卉物流服务区、岭南盆景园、苗木引种驯化示范区等。

 游览攻略
TRAVEL GUIDES

地址：花都区芙蓉大道
Address: Furong Avenue in Huadu District

It is in Shiling Town, Guangzhou, Capital of Flowers in Guangzhou aims at forging the brand of "Capital of Flowers in Guangzhou" and constructing flowers and seedlings trading center in southern China, taking the implementation of information exchange and promotion of industrial development as the focus, to strive to establish a comprehensive large flower market integrating landscaping, flower and seedlings exhibition and trade, science and technology information exchange, cultural tourism, investment invitation and agricultural eco-tourism. The major functional zone at phase I of 2000 acres can be divided into: Floral Theme Square, Seedling Trading Market, Greenhouse Flower District, Flower Wholesale Market, Household Garden Center, Garden Supplies Trading Center, Integrated Service Center, Flower Material Market, Flower Logistics Service Area, Lingnan Bonsai Garden Displaying Potted Landscape, Seedlings Introduction and Domestication Demonstration Area, etc.

GUIDE TO HUADU
花都全攻略 狮岭镇 / SHILING TOWN

👁 华严寺
HUAYAN TEMPLE

　　花都华严寺原名观音寺，始建于清宣统元年（1908年），是芙蓉山脉浩气所归，地脉雄浑，状如莲瓣，清幽典雅，身临其境，万尘俱忘。历史上信众群集，香火兴盛，近百年来几经兴废。华严万佛宝殿，全部采用印尼进口红木波罗格，52根大石柱为福建惠安优质花岗岩，最大直径为0.6米，高8米。大殿内供万尊圣像，当中本师释迦牟尼佛、大智文殊菩萨、大行普贤菩萨三尊，高6米，所有大小佛像均用紫铜铸造，面貌生动，栩栩如生。殿内壁雕"善财童子五十三参"，造型活泼可爱，色彩艳丽多姿，置身其中，备感法喜充满。殿外墙壁遍围用红木刻制的本焕长老血书字体《华严经普贤菩萨行愿品》。整座大殿古朴典雅，庄严肃穆。

　　Huadu Huayan Temple, originally named as Guanyin Temple, was built in the first year (year of 1908) of Emperor Xuantong in Qing Dynasty. Deriving from the noble spirit of Furong Mountain Range, it is characterized by magnificent geographical position in shape of lotus petals, quiet, beautiful and graceful, just like personally on the scene and forgetting the mortal world. In history, believers like the clouds brought in good attendance. It rose and fell for several times in recent hundreds of years. The Huayan Buddha Hall is made of merbau, rosewood imported from Indonesia, and 52 pieces of large stone peristeles made of high-quality granite produced in Huian of Fujian is eight meters high, with maximal diameter of 0.6 meters. There are ten thousand of icons in the hall, with Buddha Shakyamuni, Manjusri and Samantabhadra Bodhisattva in the middle, 6 meters in height. All the figures of Buddha, large or small, are vividly cast of red copper, with lifelike feature. "Fifty-three figures of Sudhana" are carved on the inner wall of the hall, characterized by lively and active shape, gorgeous and colorful. We will feel peace and pleasant when standing closely with them. The outside wall of the hall is enclosed with rosewood carved with bloodletter of Benhuan Presbyter-Huayan Classis Puxian Buddha Wishes. The overall hall is simple, unsophisticated and elegant in a solemn atmosphere.

游览攻略
TRAVEL GUIDES

地址：狮岭镇芙蓉大道西侧华藏山　　电话：020-86993026
Address: Huazang Mountain, at west side of Furong Avenue, Shiling Town
Telephone: 020-86993026

GUIDE TO HUADU
花都全攻略 狮岭镇/SHILING TOWN

花都好吃

 ### 永聚贤 Yongjuxian

永聚贤在芙蓉峰上，如端坐云间，竹林翠柏环绕其中，在此，您尽可以享受自然的静与和谐，品味心灵的闲情逸志，拓展赏心悦目的视野，进入雅致温馨的家园。
必尝推荐：农家菜为主、山水豆腐、山坑鱼、山坑螺。

Located on the Furong Peak and surrounded by bamboo forest and incense cedar, Yongjuxian is just like sitting on the clouds, where enables you to enjoy quietness and harmony of the nature, taste leisurely and carefree mood of the soul, feast your eyes on charming scenery and enter into elegant and comfortable homeland.
Recommended delicacies: featured with farm food, such as Cold Bean Curd, Shankeng Fish, Shankeng Paludina.

 地址：芙蓉度假区山顶永聚贤山庄内　订座电话：020-86853830
Address: inside Yongjuxian Mountain Villa, on the peak of Furong Tourist Resorts　Reservation hotline: 020-86853830

 ### 东成山庄 Dongcheng Mountain Villa

规模较大的山庄，车位很多，服务热情。小桥流水亭台青瓦的园林尤显精秀，环境干净整洁，古朴别致。鲩鱼很值得推荐，肉质极为细嫩鲜甜，只有低温水冷水质好的水库里才能产出。
必尝推荐：清蒸鲩鱼、客家酿豆腐、椒盐鸟、椒盐蛇。

It is a large-scale mountain villa with many parking plots and featured in warm service. Small bridge, flowing water, pavilions and grey tiles highlight the particularly exquisite and beautiful garden, clean, simple and unique. Grass carps here are highly recommended for their delicate, delicious and sweet taste, which can only be produced in reservoir with low-temperature and good water-coolong and water quality.
Recommended delicacies: Steamed Grass Carp, Hakkas Stuffed Tofu, Spicy Slat Bird and Spicy Slat Snake.

 地址：山前旅游大道瑞边村路段　订座电话：020-86993488
Address: section in the Ruibian Village, Shanqian Travel Avenue　Reservation hotline: 020-86993488

Gourmet in Huadu

方街
FANGJIE STREET

位于芙蓉嶂半山谷，里面有蓝球场、开阔活动场地、烧烤吧、茶园等，美食与娱乐功能兼备。方舟餐厅依山而建，风景赏心悦目。农家特色美食味道颇正，让人回味。

必尝推荐：水库鲩、坑螺鸡、红焖老鼠、青头鸭。

Located at the hillside of Furong Peak, there is basketball court, spacious activity space, barbecue terrace and tea garden, offering both delicacies and entertainment. The Fangzhou Canteen is constructed near the mountain for pleasant and delightful scenery. Farm special cuisines are pure and authentic, with an endless savor.
Recommended delicacies: Reservoir Grass Carp, Shankeng Paludina and Chicken, Braised mice in Brown Sauce and Aythya baeri.

  地址：芙蓉度假区山顶　　订座电话：020-86982818
Address: Mountaintop of Furong Tourist Resorts
Reservation hotline: 020-86982818

齐鲁之家山庄
MOUNTAIN VILLAGE OF QILU HOME

各类特色菜品种繁多、丰俭由人，资深名厨主理，出品精益求精，是宴客聚餐的理想地方。服务一流，信誉至上，尊贵美食享受恭您的光临。

必尝推荐：蒸鱼、香辣蟹。

Various kinds of special cuisines are enough for selection, which are cooked by famous senior chefs for constantly making better. Due to first-class service, high reputation and exquisite cuisines, it is an ideal place for entertaining guests and holding dinner party. Welcome to Mountain Village of Qilu Home.
Recommended delicacies: Steamed Fish and Spicy Crab.

  地址：狮岭镇罗仙村　　订座电话：020-86990090
Address: Luoxian Village, Shiling Town
Reservation hotline: 020-86990090

王子山森林公园
PRINCE MOUNTAIN FOREST PARK

梯面镇篇
花都全攻略
TIMIAN TOWN
OVERALL GUIDE FOR TRAVEL IN HUADU

梯面镇是广州市辖区内唯一的山区镇，位于广州市花都区北缘，是当年从北进出广州城必经之地，清朝道光年间修建花岗岩石板形成"百步梯"。"百步梯"之上为梯面，因此得名。全镇形似一棵大树，这里野生动植物种类繁多，旅游景点丰富，有王子山森林公园、紫霞山庄、高百丈风景区。这里一年四季空气清新、景色宜人。春天，山花烂漫，蜂飞蝶舞，金灿灿的油菜花惹人陶醉；夏天，青松翠柏，潺潺流水，阵阵清风沁人心田；秋天，田间小路，稻谷飘香，满目的金黄令人愉悦；冬天，暖阳微照，北风轻吹。

Located in north rim of Huadu District, Guangzhou City, it is the only town in the mountainous area under the administration of Guangzhou. Since it is the only way for passing through Guangzhou, granite slabstones were established during the reign of Emperor Daoguang in Qing Dynasty for "Hundred Stair". It becomes well-known for tread on the "Hundred Stair". The shape of the town area is like a large tree,there are a great variety of wild animals ant plants, there are lots of tourist attractions:Privins Moutain Forest Park Zixia Moutain Villa Gaobaizhang Scenic Zone.Clean and fresh air beautiful scenery are all the year round. In spring, we will be enchanted by mountain flowers in full bloom, flying and dancing honeybees and butterflies as well as golden rape flower; in summer, green pine and incense cedar, gurgling flowing water and refreshing breeze will moisten our hearts. In autumn, it will feast our eyes on fragrance of fresh grain drifted on the breeze from path in the field, with a golden scene. In winter, we will be pleasant by warm sunshine and gentle north wind.

数字看梯面
FIGURES OF TIMIAN TWON

总面积91.2平方千米，山林面积11.8万亩，占总面积的88%。
The gross area is 91.2 square kilometers and the area of mountain forest is 118000 Mu, accounting for 88% of the gross area.

GUIDE TO HUADU
花都全攻略 梯面镇/TIMIAN TOWN

王子山森林公园
PRINCE MOUNTAIN FOREST PARK

　　王子山森林公园是广州少有的原生态氧吧，由芙蓉度假区和王子山林场两部分组成。王子山林木苍翠，山势巍峨，公园北部的牙英山海拔581.2米，为全区最高峰，相邻的王子山海拔571.9米。空气质量经省林业勘查院测定为"最清洁空气"，负离子含量高，树高林密，溪水清澈，还隐藏有山洞秘道，是寻幽探秘的好地方。森林公园中至今仍有野生动物频繁出没，常见的有野猪、果狸、山兔、山鸡等。王子山的植物群落构成类型很多，先后发现了10多株国家一级保护植物桫椤，其植物群落多为多层林，中草药材丰富，有药源品种100多个。

The Province-Level Prince Mountain Forest Park is one of the few original ecological oxygen bars in Guangzhou, consisting of Furong Tourist Resorts and Prince Mountain Forest Farm, where forests are lush and verdant and mountains are tall and rugged. Yaying Mountain in the north of the park with 581.2 meters elevation is the peak of the overall area, and the adjacent Prince Mountain is 571.9 meters high. Air quality is determined by Provincial Forestry Exploration Institution as "the cleanest air" for high content of anion. It is really a good place for exploring quiet and peaceful environment and secretes for tall trees, dense forests, clear stream and hidden cave and secrete passageway. Wild animals still frequently appear and disappear in the forest park and wild boar, paguma larvata, hare and pheasant are common to see. The phytoecommunity in Prince Mountain are consisted of various types and dozens of spinulose tree fern, first-grade state protection plant, have been discovered one after another, whose phytoecommunity is polylayer forest, together with abundant Chinese medicinal herbs and over one hundred varieties of medicinal herbs resources.

★ 推荐看点
RECOMMENDED VIEWING FOCUS

王子山海拔高571.9米，有"花都白云山"之称，王子山森林公园负离子含量120000个/cm^3，登临山顶后如人处仙境般的云雾。飞瀑流泉遍布其间，沿途峰回路转，一弯一景，云雾缭绕，据说区内还隐藏有山洞秘道，是寻幽探秘的好地方。

Prince Mountain, 571.9m high, is renowned as "Baiyun Mountain of Huadu". The content of anion in the Prince Mountain Forest Park is 120000 per cubic centimeter, hence, you will feel personally stand in the fairyland after arriving at the mountaintop and surrounded by cloud and mist. Waterfall and flowing spring can be found everywhere, with path winding along mountain ridges. Every corner could structure scenery surrounded by clouds and mist. It is said that there are hidden caves and secrete passageways which deserves the exploration of quiet and peaceful environment and secretes.

游览攻略
TRAVEL GUIDES

地址：花都区梯面镇西坑村
电话：020-86759399　　　门票：30元
公交线：在花都客运站乘65路车到达梯面镇，再转乘摩托车即可。或从广州机场路电车总站乘坐704路公交车直达梯面镇。
自驾线1：沿106国道北上，经花山镇到梯面镇中心，然后顺指示牌自东往西进入王子山。
自驾线2：从芙蓉门经芙蓉度假区内的主干道，进入王子山。

Address: Xikeng Village, Timian Town, Huadu District
Telephone: 020-86759399　　　Entrance ticket: 30 Yuan
Bus route:take the Huadu No.65 bus at Huadu Passenger Station to Timian Town,trun to by a motor bicycle,or Take the No.704 bus at Guangzhou Airport Road Station to Timian Town.
Self-driving route 1: go north along G106 through Huashan Town to Timian Town Center and drive to Prince Mountain from east to west as instructed by indicator.
Self-driving route 2: drive from Furong Gate to main stem of Furong Tourist Resorts to enter into Prince Mountain.

GUIDE TO HUADU
花 都 全 攻 略　梯面镇 / TIMIAN TOWN

👁 高百丈风景区
Gaobaizhang Scenic Zone

　　高百丈风景区自古以来就是花县八景之一，称为"百丈晴峦"，以遍布奇石怪岩闻名。高百丈最巨大的岩石，石壁面对东南，160多年前有一位和尚在这石壁上刻了"百丈晴峦"四个大字，现在仍然字迹清楚，苍劲有力。

Gaobaizhang is one of "Eight Scenery in Huaxian County" since ancient times and named as "Bai Zhang Qing Luan" (Mountains in Sunny Weather for Hundreds of Meters) for rare stones and strange rocks everywhere. The cliff of the largest rock is facing the southeast and over one hundred and sixty years ago, a monk carved four characters of "Bai Zhang Qing Luan" on the cliff. So far, the handwriting is still clear, vigorous and forceful.

推荐看点
Recommended Viewing Focus

高百丈风景区中有众多著名的奇石，如"雷公劈石"、"了哥髻"、"石棺材"、"醉翁石"等，都是以形似而得名。这里树林茂盛，山泉不竭，形成多级小瀑布，别有一番风光。

There are plenty of rate stones in the Gaobaizhang Scenic zone, such as "Thunder God Splitting Stone" "Comb of Crested Myna" "Stone Coffin" "Stone of Drinker", all of which are named for similar shape. The scenic zone is characterized by lush forests, and inexhaustible mountain spring, forming small falls at different levels, which brings us a specific interest.

游览攻略
Travel Guides

地址：花都区梯面镇联民村
公交线：在花都客运站乘65路车到达梯面镇，再转乘摩托车即可。或从广州机场路电车总站乘坐704路公交车直达梯面镇。
自驾线：沿106国道北上，经花山镇到梯面镇，顺着指示牌到高百丈。

Address: Lianmin Village, Timian Town, Huadu District
Bus route:Take the Huadu No.65 bus at Huadu Passenger Station to Timian Town,turn to by a motor bicycle,or take the No.704 bus at Guangzhou Airport Road Station to Timian Town.
Self-driving route:Move north along S106,Pass by Huashan Town to Timian Town, by the instruction of guidepost to Gaobaizhang Scenic Zone.

GUIDE TO HUADU

花都全攻略 梯面镇 / TIMIAN TOWN

美丽乡村红山村
BEAUTIFUL VILLAGE- HONGSHAN VILLAGE

红山村位于梯面镇西北部，是王子山脚下一座小村庄，总面积14.3平方千米。2月中到3月初可到红山村赏油菜花，这里有大片金黄的油菜花，游人在田间嬉戏，蜜蜂在花间飞舞，既可沿着小路溯溪徒步而上，也可以踩着自行车享受春风拂面的感觉。2013年1月成为首个美丽乡村国家AAA级旅游景区。

Located in the northwest of Timian Town, Hongshan Village is a small village at the foot of Prince Mountain. Hongshan Village covers an area of 14.3 square kilometers. Every year in the middle third of February and early March, visitors can go to Hongshan Village to appreciate a cloud of golden rape flowers and play with each other among fields. With honeybees flying among blossom, you can wade along footpath on foot or ride a bike to enjoy a spring breeze stroking the face. It became the first beautiful countryside national AAA level scenic spots on January, 2013.

 推荐看点
RECOMMENDED VIEWING FOCUS

　　红山村内还有深浅二谷,均有瀑布,有"石上飞瀑"、"深谷幽峡"、"浅谷幽深"、"红谷幽情"等景点。红山村拥有自行车出租服务,可租自行车亲身体验美丽乡村的农家色彩。

There are two velleys, one deep valley and another shallow one, in the Hongshan Village, with falls respectively, famous for "Waterfall on the Stone" "Quiet Gorge in Deep Valley" "Deep and Quiet Shallow Valley" "Deep and Quiet Red Valley" and other scenery. Hongshan Village offers bicycle rental service, enabling visitors to experience farmhouse atmosphere of beautiful village by bike.

 游览攻略
TRAVEL GUIDES

自驾线:广州—上机场高速—花都出口—106国道—向梯面镇方向行驶,注意不要驶入花都市区。经花山镇进入梯面镇。
背包族线路:乘车到花都客运站—转乘"新华站—梯面站"65号专线车

Self-driving: Guangzhou- drive to Airport Expressway- Huadu Exit- G106- drive toward Timian Twon instead of Huadu Downtown, and enter into Timian Town after passing through Huashan Town.
For Backpacker: go to Huadu Passenger Station by bus- transfer to No.65 Special Shuttle Bus of "Xinhua Station- Timian Station".

GUIDE TO HUADU
花都全攻略 梯面镇 / TIMIAN TOWN

紫霞山庄
ZIXIA MOUNTAIN VILLA

　　位于梯面镇东北部的紫霞山庄，自然环境优雅，位于群山环抱之中，周边瀑布、溪流环绕，是修心养性的好去处。关于紫霞山庄，当地人还流传着这样的说法：很久很久以前，紫霞山庄中的"大雄宝殿"前有一座"仙人桥"，这道"桥"原来是仙人用大榕树的一条树根"建造"起来的，可见，山庄中仙气浓郁，是仙人雅士们愿意待的地方。

Located in the northeast part of Timian Town, Zixia Mountain Villa is famous for elegant natural environment. Embraced by mountains, falls and stream, it is a good place for cultivating our mind and improving our character. There is an old saying passed by local people: once upon a time, there was a "Fairy Bridge" in front of "Main shrine Hall" of Zixia Mountain Villa, which was "constructed" by immortal with a piece of tree stump of ficus microcarpa. It is thus clear that immortal and refined scholars are willing to stay in Zixia Mountain Villa for rich magic ambience.

推荐看点
RECOMMENDED VIEWING FOCUS

水流石，卧佛岭，老虎洞，仙人桥，藏珠台，贵子亭，万福塔。

Stone of Flowing Water, Mountain Range of Reclining Buddha, Tiger Cave, Fairy Bridge, Pearl Collection Platform, Guizi Pavilion and Blissful Tower.

游览攻略
TRAVEL GUIDES

地址：梯面镇五联村　　　电话：020-86851800　　　门票：5元
自驾线：广州—机场高速—花都出口—106国道—梯面方向行驶—花山镇—梯面镇。
背包族线路：乘车到花都客运站，转乘"新华站—梯面站"65号专线车。

Address: Wulian Village, Timian Town
Telephone: 020-86851800
Entrance ticket: 5 Yuan
Self-driving route: Guangzhou- Airport Expressway- G 106- drive toward Timian Town- Huashan Town - enter into Timian Town.
For Backpacker: go to Huadu Passenger Station by bus- transfer to No. 65 Special Shuttle Bus of "Xinhua Station- Timian Station".

GUIDE TO HUADU
花都全攻略 梯面镇 / TIMIAN TOWN

花都好吃

🍴 红山乡村酒店
RURAL HOTEL OF HONGSHAN VILLAGE

红山乡村酒店位于被誉为广州最美乡村的梯面镇红山村内，是一家融入乡村历史文化、突出自然乡土气息和风情的高档次乡村酒店。配有住宿、餐饮、会议室、篮球场、羽毛球场、烧烤场、绿道单车等设施。

Located in Hongshan Village of Timian Town, the most beautiful village in Guangzhou, Hongshan Rural Hotel is a high-end rural hotel integrating rural history and culture, and highlighting natural and rural flavor and style. And it is also equipped with accommodation, catering, meeting room, basketball court, badminton court, barbecue ground, greenway bicycle and other facilities.

 地址：梯面镇红山村　订座电话：020-37700388/37700828
Address: Hongshan Village, Timian Town
Reservation hotline: 020-37700388,37700828

🍴 雁鹰山庄
YANYING MOUNTAIN VILLA

雁鹰山庄位于花都区梯面镇王子山生态旅游区的秀丽山林之间。这里的负离子含量极高，空气清新指数和山泉指数广州地区最高。

Located in beautiful mountains forests of Prince Ecological Tourism Area, Timian Town, Huadu District, the Yanying Mountain Villa boasts high content of anion and highest indexes of fresh air and mountain spring in the Guangzhou area.

 地址：花都区梯面镇　订座电话：020-86782833
Address: Timian Town, Huadu District
Reservation hotline: 020-86782833

Gourmet in Huadu

瑞记农家庄
Rui's Farm Village

必尝推荐：风沙鸡、客家卤肉等农家招牌菜。
Recommended delicacies: Sauced Chicken and Hakka Belly Pork are farm specialties.

 地址：梯面镇红山村　　订座电话：13682221790
Address: Hongshan Village, Timian Town
Reservation hotline: 13682221790

砂糖桔山庄
Sugar Orange Mountain Villa

必尝推荐：特色秘制山猪肉，野生石蛤。
Recommended delicacies: Special Boar Meat and Wild Chukar.

 地址：梯面镇联丰村　　订座电话：18818397023

高望饭店
Gaowang Hotel

必尝推荐：碌鹅、鲜鱼豆腐汤、咸鱼炆花腩。
Recommended delicacies: Boiled Goose, Fresh Fish and Beancurd Soup, and Braised Salted Fish and Streaky Pork.

 地址：梯面镇联民村　　订座电话：020-86781032
Address: Lianmin Village, Timian Town
Reservation hotline: 020-86781032

联兴山庄
Lianxing Mountain Village

必尝推荐：正宗水库鱼、山水豆腐、梯面烧肉、山水豆腐。
Recommended delicacies: Authentic Reservoir Fish, Timian Braised Pork and Cold Bean Curd.

 地址：梯面镇联民村　　订座电话：020-86851555
Address: Lianmin Village, Timian Town
Reservation hotline: 020-86851555

友和酒店
Youhe Hotel

必尝推荐：石木鱼仔、砵仔鹅、菜干煲、土鸡田螺煲。
Recommended delicacies: Stone Bowl Cuttlefish, Cooked Goose in Clay Pot, Dry Vegetable in Clay Pot, Local Chicken and Viviparus in Clay Pot.

 地址：梯面市场侧边　　订座电话：020-86782888
Address: On the side of Timian Market
Reservation hotline: 020-86782888

花都湖
PRINCE MOUNTAIN FOREST PARK

雅瑶镇篇
花都全攻略
YAYAO TOWN
OVERALL GUIDE FOR TRAVEL IN HUADU

雅瑶镇位于广州市花都区的南部，京广铁路、武广高速铁路从境内穿过。雅瑶镇的名片是"中国音响之都"。

Yayao Town is in the southern part of Huadu District, Guangzhou, Beijing-Guangzhou Railway and Wuhan-Guangzhou High-Speed Railway pass the Yayao Town, which is named as "Capital of Speaker Box in China".

数字看雅瑶
FIGURES OF YAYAO

雅瑶镇总面积8.64平方千米。

The gross area of Yayao Town is 8.64 square kilometers

GUIDE TO HUADU
花都全攻略 雅瑶镇 / YAYAO TOWN

👁 花都湖
HUADU LAKE

花都湖，东起铁山河的河口，西至京广铁路路桥段，全长约6.68千米，占地总面积约2.83平方千米。正在建设的花都湖公园在设计上充分体现亲水性，力求保留历史特色风貌，融入本土文化元素，它的开发建设将有力带动周边约23平方千米的南部新城区发展，从而打通花都主城区与广州中心城区的连接，促进广州"北优"战略进程。

Huadu Lake approximately 6.68 kilometers in total starts from river mouth in the Tieshan River to the bridge of Beijing-Guangzhou Railway in the west, covering approximately 2.83 square kilometers of gross area in total. The Huadu Lake Park under construction sufficiently demonstrates water-enjoyable design to strive to retain historical characteristics and style and integrate local cultural elements. The development of Huadu Lake Park will forcefully drive the development of surrounding new urban district approximately 23 square kilometers in the south, so as to connect main urban area of Huadu with central urban area of Guangzhou, for the purpose of the strategic process of "North Optimization" of Guangzhou.

✏ 游览攻略
TRAVEL GUIDES

地址： 雅瑶镇滨湖路东侧
自驾线： 机场高速—花都出口—迎宾大道—镜湖大道北—雅瑶中路—114省道—滨湖路。

Address: East Side of Binhu Road, Yayao Town
Self-driving route: Airport Expressway- Huadu Exit- Yinbin Avenue- north of Jinghu Avenue- Yayao Middle Road- S 114- Binhua Road.

椰林2号海鲜码头
No.2 Seafood Wharf in Coconut Grove

环境宽敞有特色，风景怡人，海鲜新鲜且很平价，鱼肉尤为细嫩鲜美。
必尝推荐：锡纸焗鱼、芝士龙虾。

The environment is specious and unique, with pleasing scenery. The seafood is fresh in fair price, with delicate and delicious fish.
Recommended delicacies: Baked Fish with Silver Paper, and Cheese Lobster.

地址：雅瑶镇雅瑶中路1号　　订座电话：020-86823112
Address: No.1 Yayao Middle Road, Yayao Town
Reservation hotline: 020-86823112

芬姐蒸鸡
Steamed Chicken by Sister Fen

装修很有乡村味道，铜盘蒸鸡极受推崇，正宗的广西水库鱼肉质结实鲜美。
必尝推荐：铜盘蒸鸡、豉汁鱼腩。

The decoration is full of rural style, Steamed Chicken in Copper Disc, and authentic Guangxi Reservoir Fish is strong and delicious.
Recommended delicacies: Steamed Chicken in Copper Disc, and Fish Belly with Bean Sauce.

地址：雅瑶镇邝家庄路口　　订座电话：020-86164978
Address: Road junction of Kuangjiazhuang Thorpe, Yayao Town
Reservation hotline: 020-86164978

花都广场夜景
NIGHT SCENE OF HUADU PLAZA

线路推荐 特色主题

TOURISM LINE PROMOTION

OVERALL GUIDE FOR TRAVEL IN HUADU

TOURISM LINE PROMOTION 特色主题线路推荐

畅游花都，你还可以这样玩！

🚗 养生休闲游 TOURISM OF HEALTH PROMOTION AND LEISURE

花都喜来登度假酒店、九龙湖公主酒店、广东王子山森林公园、广州高百丈森林公园、广州九湾潭森林公园、花都蟾蜍石森林公园、花都福源水森林公园及花都丫髻岭森林公园等六个森林公园。

There are six forest parks: Huadu Sheraton Resort Hotel, Dragon Lake Princess Hotel, Guangdong Prince Mountain Forest Park, Guangzhou Gaobaizhang Forest Park, Guangzhou Jiuwan Pond Reservoir Forest Park, Huadu Toadstone Forest Park, Huadu Fuyuan Water Forest Park and Huadu Yajiling Forest Park.

🚗 文化体验游 TOURISM OF CULTURAL EXPERIENCE

资政大夫祠古建筑群、盘古王庙、炭步镇塱头村民居古建筑群、高溪村欧阳庄民居古建筑群等。

Ancient Architectural Complex of Ancestral Temple of Zizheng Senior Official (an official position in Yuan, Ming and Qing Dynasties), Temple of King of Pan Gu, Ancient Residential Architectural Complex of Langtou Villager of Tanbu Town, and Ancient Residential Architectural Complex of Ouyangzhuang Thorpe of Goxi Village

🚗 主题公园游 TOURISM OF THEME PARK

石头记矿物园、香草世界、宝桑园、广州花卉之都、瑞岭盆景村等。

Shitouji Mineral Park (famous for stones), Vanilla World, Baoshang Park, Capital of Flower in Guangzhou, Ruiling Village (famous for potted landscape)

🚗 花都绿道游 TOURISM OF HUAD GREENWAY

山清水秀的芙蓉嶂水库是广东省4号绿道的起点，途经大布河、芙蓉市场、洪秀全水库、杨赤公路、狮岭镇草弄村、红棉大道，结束于天马河，全程约30千米。广东省2号绿道，主线长约470千米，在花都境内约15千米，沿流溪河分布。

The Furong Peak Reservoir with picturesque scenery is the starting point of No.4 Greenway of Guangdong Province, which passes through Dabu River, Furong Market, Hongxiuquan Reservoir, Yangchi Highway, Caonong Village of Shiling Town, Hongmian Avenue to Tianma River, approximately 30 kilometers in total. The No.2 Greenway of Guangdong Province is approximately 470 kilometers in total, and about 15 kilometers of No.2 Greenway passes through Huadu along stream and river.

🚗 宗教修持游 TOURISM OF RELIGIOUS CULTIVATION

华严寺、圆玄道观、盘王古庙、八角庙、紫霞山庄等宗教圣地。

Huayan Temple, Yuanxuan Taoist Temple, Temple of King of Pan Gu, Bajiao Temple, Zixia Mountain Villa and other holy lands of religion.

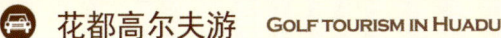

Entertainment recommended for travel in Huadu

🚗 花都高尔夫游 GOLF TOURISM IN HUADU

广州九龙湖高尔夫球会，位于花都区花东镇九龙湖国际社区；广州风神高尔夫球会，位于花都区风神大道转入花港大道；广州美林湖国际乡村俱乐部，位于花都区山前大道美林湖国际社区内。

Guangzhou Dragon Lake Golf Club located in Dragon Lake International Community, Huadong Town, Huadu District; Guangzhou Foison Golf Club located in the Fengshen Avenue tuning to Huagang Avenue in Huadu District; Guangzhou Mayland International Rural Club, located in Mayland International Community, Shanqian Avenue, Huadu District.

🚗 工业产业游 INDUSTRIAL TOURISM

皮具皮革旅游、汽车产业基地旅游、珠宝工业旅游。

Tourism of leatherware and leather, tourism of automobile industrial base, and tourism of jewelry industry

🚗 特产物候游 TOURISM OF SPECIALTY AND FESTIVALS

花都特色物产丰富，每年会举办多种节庆游，如炭步芋头节、枇杷节、桑果节、香草文化节等。

Huadu not only produces abundant specialties, but also hold various festivals tourism per year, such as Tanbu Taro Festival, Loquat Festival, Mulberry Festival, Vanilla Cultural Festival etc.

🚗 民俗节庆游 TOURISM OF FOLK FESTIVALS

花都民俗活动特别丰富，保存良好，到花都参与民俗节庆旅游，可穿越时空，聆听亘古声音。

The well preserved Huadu folk activities are abundant and unique. Go to Huadu and participate in the tourism of folk festival, and you can listen to the sound from ancient times through the veil of time.

🚗 婚纱摄影游 TOURISM OF WEDDING PHOTOGRAPHING

花都山水秀美，众多豪华主题度假酒店，如九龙湖公主酒店、花都喜来登度假酒店、王子山、芙蓉度假区等均为婚纱摄影外景的最佳选择。

In addition to beautiful landscape, Huad boasts plenty luxury theme resorts, for example, Dragon Lake Princess Hotel, Huadu Sheraton Holiday Hotel, Prince Mountain and Furong Tourist Resorts are all best choices for outdoor scene of wedding photo.

🚗 空港转机游 TOURISM OF AIRPORT TRANSFER

花都具有毗邻机场的地理优势，客人在国际机场转机6小时以上的，都可发展"转机游"项目，广州，仅有花都具备这样的条件。

By virtue of the geographical advantage of adjacent to airport, visitors waiting in the International Airport for airplane transfer for over six hours could turn to "tourism of airport transfer". Only Huadu District possesses such condition throughout Guangzhou.

新世纪酒店
NEW CENTURY HOTEL

介绍 花都酒店

HOTEL INTRODUTION OF HUADU

HOTEL INTRODUTION OF HUADU 花都酒店介绍

🄷 九龙湖公主酒店 DRAGON LAKE PRINCESS HOTEL ★★★★★

区内首家由国家旅游局授牌的五星级酒店，位于九龙湖度假区内，由15栋不同风格的欧式建筑群组成，是国内顶级度假圣地，并被誉为广州奢华客房的典范。拥有15间不同规格的会议厅，是广州首屈一指的商务活动场地。配备完善一流的康体娱乐设施，是休闲健身的绝佳场所。

Located in the Dragon Lake Resorts, Dragon Lake Princess Hotel is the first five-star hotel approved by National Tourism Administration nationwide, consisting of 15 European-style architectural complexes in different styles. It is the top-level domestic resort praised as the model of luxurious guest room in Guangzhou. Possessing 15 conference halls in different specifications, it is the first of all commercial activity yards in Guangzhou, as well as the best place for recreation and body-building with complete and first-class gymnasium and recreational facilities.

 地址：花东镇九龙湖社区　　电话：020-36908888
Address: Dragon Lake Community, Huadong Town　　Telephone: 020-36908888

🄷 新世纪酒店 NEW CENTURY HOTEL ★★★★

区内唯一由国家旅游局授牌的四星级旅游酒店，按五星级标准装修，是花都标志性酒店。地理位置极为优越，机场穿梭BUS和港澳专线BUS均在此设专用候车站。配套先进完善，拥有网球场、健身房、室外游泳池等娱乐设施。

New Century Hotel, a landmark in Huadu, is the only four-star tourist hotel approved by National Tourism Administration nationwide and decorated in line with five-star standard. It boasts extremely superior geographic position that dedicated departure terminal is established by Airport Shuttle BUS and Special Hong Kong and Macao BUS Route. It is equipped with advanced and complete facilities, together with tennis court, gymnasium, outdoor swimming pool and other recreational facilities.

 地址：新华街秀全大道43号　　电话：020-36858888
Address: No. 43 Xiuquan Avenue, Xinhua Street　　Telephone: 020-36858888

合兴大酒店 Hexing Hotel ★★★

集商旅、会议、KTV、康乐和美食于一体的综合性酒店，设有旋转餐厅、游泳池、篮球场、乒乓球室和大型停车场等。主楼高25层，是花都区又一标志性建筑。

Hexing Hotel integrates business travel, conference, KTV, recreation and delicacies as a whole, equipped with revolving restaurant, swimming pool, basketball court, table tennis room and large parking lot, etc. The main building has 25 storeys, making it another landmark of Huadu District.

地址：新华街建设北路213号　　电话：020-36996888
Address: No. 213 Jianshe North Road, Xinhua Street　　Telephone: 020-36996888

嘉尔登酒店 Jia Erden Hotel ★★★

欧陆风格综合大酒店。酒店设有中西餐厅、夜总会、游艺、桑拿沐足休、健身、台球室、室外游泳池等康乐设施，是旅客在商旅中享受休闲时刻的快乐驿站。

Jia Erden Hotel is an integrated hotel in European style equipped with Chinese and Western restaurants, nightclubs, entertainment, sauna, footbath, recreation, body-building, billiard parlor, outdoor swimming pool and other recreational facilities, becoming the recreation area for visitors enjoying leisure time during business travel.

地址：新华街曙光大道30号　　电话：020-36802888
Address: No. 30 Shuguang Avenue, Xinhua Street　　Telephone: 020-36802888

荣威大酒店 Rongwei Hotel ★★★

位于区心脏地带迎宾大道上。宴会厅、各类多功能会议厅、西餐咖啡厅、豪华KTV、健身、桑拿、沐足、美发美容等康乐设施一应俱全。

Rongwei Hotel is a tourist hotel in the Yinbing Avenue, the headland of urban downtown. It isequipped with banquet hall, various types of multi-functional conference halls, Western-style food restaurant and coffee house, luxury KTV, body-building, sauna, footbath, hairdressing, cosmetology and other recreational facilities.

地址：新华街迎宾大道13号　　电话：020-86889888
Address: No. 13 Yinbing Avenue, Xinhua Street　　Telephone: 020-86889888

HOTEL INTRODUTION OF HUADU 花都酒店介绍

华悦酒店 HUAYUE HOTEL ★★★

装修设计幽雅舒适，内部设施先进，配备齐全。酒店内设中餐厅、咖啡厅、客房送餐服务。商场、银行、汽车站近在咫尺，距广州白云国际机场仅需15分钟。

The hotle is decorated with elegant and comfortable design, equipped with complete advanced in-house facility. The hotel offers Chinese restaurant, coffee house and room service. The transportation is absolutely convenient that shopping mall, bank and bus station are near at hand, only 15 minutes away from Guangzhou Baiyun International Airport.

 地址：新华街建设北路128号　　电话：020-36883888
Address: No. 128 Jianshe North Road, Xinhua Street　　Telephone: 020-36883888

亨利商务酒店 HENRY BUSINESS HOTEL ★★★

位于宝华路繁盛商业区核心地带，是由美国太平洋环球投资有限公司独资组建的首家国内商务型连锁酒店。建筑设计与众不同，前卫新颖的室内空间，从不同角度都可拥有良好视觉感。全新概念的泰国餐厅誉享花都。

Located at the center of Fansheng Commercial District in Baohua Road, Henry Business Hotel is the first domestic business chain hotel solely invested and established by American Pacific Global Investment Company. Different architectural design as well as advanced and novel interior space enables you to feel sound visual sense in different angles. The Thai Restaurant with bran-new concept is renowned throughout the overall Huadu.

 地址：新华街宝华路26号　　电话：020-36812888
Address: No. 26 Baohua Road, Xinhua Street　　Telephone: 020-36812888

合成大酒店 HECHENG HOTEL ★★★

商务休闲酒店，紧邻狮岭国际皮具皮革城，新白云国际机场近在咫尺。独特欧陆风格，美食、购物、休闲和商务等配套设施齐全，使商务居停的您备感轻松自如！

It is a business and leisure hotel close to Shiling International Leatherware and Leather City and the new Baiyun International Airport. Completely equipped with delicacies, shopping, leisure and business in European style, it will make it very easy for your business stay.

 地址：狮岭镇狮岭大道中1号　　电话：020-36919168
Address: No.1 Shiling Avenue Central, Shiling Town　　Telephone: 020-36919168

锦都商务大酒店 JINDU BUSINESS HOTEL ★★★

专业商务会议酒店。优越的地理环境，优质的客房服务，一流的会议、康体、娱乐配套，专业的管理，是开展商务、会议、宴请、休闲的理想首选场所。

It is a professional business conference hotel. By virtue of superior geographical environment, high-quality housekeeping service, first-class supporting conference, recreation, entertainment and specialized management, it is the first ideal place for business, conference, dinner party and leisure.

 地址：新华街滨湖路1号　　电话：020-86819999
Address: No. 1 Binhu Road, Xinhua Street　　Telephone: 020-86819999

京华酒店 JINGHUA HOTEL ★★★

地处广州北站商业区，旅游交通便利。装修温馨舒适，优雅时尚。并提供送餐、洗衣、叫醒、医疗支援、商务会议、旅游、前台贵重物品保险柜、汽车租赁等特色服务，是出差、全家度假的理想选择。

Located in the commercial district in the Guangzhou North Station, the tourist communication is quite convenient. The elegant and fashionable hotel with warm and comfortable decoration will offer such special services as room service, laundry, morning call, medical support, business conference, tourist, valuables safe case in the reception and automobile leasing and renting, becoming an ideal choice for business trip and family holiday.

 地址：新华街云山大道55号　　电话：020-36810333
Address: No.55 Yunshan Avenue, Xinhua Street　　Telephone: 020-36810333

凯利登大酒店 GRAND KINGDOM HOTEL ★★★

三星级标准豪华商务酒店，位于区政府旁，地理位置极优越。集客房、餐饮、商务、娱乐休闲于一体，设施完善，尊贵气派，服务周密。1至3楼设有不夜天中式酒楼，4至6楼为顶级装修KTV。

It is a three-star standard luxury business hotel located beside District Government, enjoying extremely superior geographic location. The hotel integrates guest rooms, catering, business, entertainment and leisure as a whole, equipped with complete facilities for noble and thoughtful service. The Sleepless Chinese Restaurant is established from first storey to third storey and a KTV decorated in top level is set from fourth to sixth storey.

 地址：新华街天贵路92号　　电话：020-36887888
Address: No. 92 Tiangui Road, Xinhua Street　　Telephone: 020-36887888

HOTEL INTRODUTION OF HUADU 花都酒店介绍

丽美大酒店 LIMEI HOTEL ★★★

涉外三星级酒店，位于繁华的商业大道，建筑雄伟，装潢华贵，旅行交通十分便利。特设商务楼层、丽美轩中餐厅、龙腾阁中餐厅及十楼西餐厅。交通便捷，是理想的旅游、商务下榻地。

It is a foreign three-star hotel located in prosperous Shangye Avenue, famous for grand and magnificent architecture, luxury decoration and convenient tourist communication. It is absolutely an ideal place for tourists and businessmen to stay on the special commercial floor, at Limeixuan Chinese Restaurant, Longtengge Chinese Restaurant and Western Restaurant in the tenth storey, along with convenient transportation.

地址：新华街商业大道53号　　电话：020-86819888
Address: No. 53 Shangye Avenue, Xinhua Street　　Telephone: 020-86819888

金湖大酒店 GOLDEN LAKE HOTEL ★★★

位于秀全公园湖畔，地处旺中带静的政要中心，集餐厅、客房、会议室、康体、娱乐设施于一体。从客房可俯瞰花都秀全公园的秀丽风光，让您在繁忙都市生活中享受轻松心情。

Located at the lakeside of Xiuquan Park and quiet political center of prosperous position, it is a hotel integrates Chinese restaurant, guest room, conference room, gymnasium and recreational facilities.

地址：新华街建设路4号　　电话：020-86808100
Address: No. 4 Jianshe Road, Xinhua Street　　Telephone: 020-86808100

金之鼎酒店 GOLDING HOTEL ★★★

位于花都区新华街滨湖路25号。是一家挂牌3星级酒店，客房超大舒适，是一座集客房、餐饮、夜总会、桑拿、商务中心和免费停车场设施于一体的挂牌三星级酒店。

Located in No. 25 Binhu Road, Xinhua Street, Huadu District, it is a three-star listing hotel integrating guest room, catering, night club, sauna, business center, free parking lot and other facilities, characterized by oversized and comfortable guest rooms.

地址：新华街天贵南路（横潭公园旁）　　电话：020-86801666
Address: Tiangui South Road, Xinhua Street (Hengtan Park)　　Telephone: 020-86801666

友田酒店 YOUTIAN HOTEL ★★★

　　按照三星级标准兴建的商务型酒店，毗邻狮岭皮革城，周围有银行、饮食及大型购物商场，交通便利。设有洗衣、票务、商务、健身、KTV、夜总会等配套设施。

It is a business hotel constructed as the standard of three-star, close to Shiling Leather City. Surrounded by bank, catering and large shopping mall, it is famous for convenient transportation. In addition, it is equipped with such supporting facilities as laundry, ticket business, business affairs, body-building, KTV and night club.

 地址：狮岭镇盘古路与东升路交界友田城　　电话：020-22689999
Address: Youtiancheng, the border between Pangu Road and Dongsheng Road, Shiling Town
Telephone: 020-22689999

中华酒店 ZHONGHUA HOTEL ★★★

　　集饮食、客房于一体的三星级涉外酒店，装修豪华，气派非凡。座落于繁华商业区中心，设有中餐厅，客房全部配有港澳、国内长途、迷你吧等多项服务设施。

Zhonghua Hotel is a foreign three-star hotel. In addition to the integration of catering and guest room, it is characterized by luxurious decoration in noble and special style. Located in the center of prosperous commercial district, it offers Chinese restaurant and guest rooms completely equipped with Hongkong, Macao and domestic long-distance telephone, mini bar and other service facilities.

 地址：新华街站前路33号　　电话：020-86822222
Address: No. 33 Zhanqian Road, Xinhua Street　　Telephone: 020-86822222

新凤凰酒店 NEW PHOENIX HOTEL ★★★

　　按照国家三星级兴建的涉外旅游酒店，邻近市中心、火车站、白云国际机场。集旅业、商务会议、饮食娱乐、休闲于一体，提供商务中心、租车、票务、洗衣、按摩、美容美发、机场接机等多项服务。

It is a foreign tourist hotel built according to national three-star level standard, close to urban downtown, railway station and Baiyun International Airport. In addition to the integration of tourist trade, business conference, catering, entertainment and leisure, it offers various services, such as commercial center, automobile leasing and renting, ticket business, laundry, massage, cosmetology, hairdressing and airport pick-up.

 地址：新华街迎宾大道大华2路38号　　电话：020-86966222
Address: No. 38 Dahua, Dahua Second Road, Yinbing Avenue, Xinhua Street
Telephone: 020-86966222

HOTEL INTRODUTION OF HUADU 花都酒店介绍

牡丹大酒店 PEONY HOTEL ★★★

涉外商务酒店，位于花都中心地带，距新白云国际机场及广州火车北站5分钟车程。拥有花都首家韩式全自动升降停车场，配备中餐厅、康体中心桑拿浴及豪华KTV。

It is a foreign business hotel and is located at the center of Huadu, with a 5-minute drive away from New Baiyun International Airport and Guangzhou North Railway Station. It boasts first Korean full-automatic stereoscopic parking lot in the Huadu, equipped with Chinese restaurant, health center, sauna and luxury KTV.

地址：新华街站前路34号　　　电话：020-36807888
Address: No. 34, Zhanqian Road, Xinhua Street　　　Telephone: 020-36807888

丽湾酒店 LIWAN HOTEL ★★★

丽湾酒店位于机场大道北出口；距广州市中心仅20分钟车程。现拥有各式豪华客房120多间，多功能会议厅可容纳100多人，并配有大型的停车场，提供极具特色的粤式佳肴，是八方来宾商务、旅游、休闲和转机的首选伴侣。

Liwan Hotel is located in the north exit of Airport Avenue. And it will take only 20 minutes to Guangzhou. Now it has over 120 rooms of all types of deluxe guestrooms, and its multifunctional conference hall can accommodate over 100 people, and it has a large-scale parking lot as well as characteristic Cantonese dishes , so that it is your first choice for business, tourism, leisure and plane change.

地址：花都区花东镇机场大道北出口　总机：86755477　传真：86766606
Address: North exit of Airport Avenue, Huadong Town, Huadu District
Switchboard: 86755477　　Fax: 86766606

合景喜来登度假酒店 ZENTH INTERNATIONAL HOTEL

位于花都区山前旅游大道东，毗邻九龙湖度假区，距离市中心四十五分钟车程、白云国际机场仅二十五分钟车程。拥有98间恬静湖景以及深邃山林风光的精美客房、套房及独立别墅，中西餐厅、多功能厅、宴会大厅、会议中心、健身中心、儿童俱乐部、户外泳池、水疗、有机农场等设施一应俱全，是商务会议、休闲度假的理想场地。

Located in Huadu District Shanqian Tourist Road East,adjacented to the Dragon Lake Resorts. It taks 25 minutes from the city center and 25 minutes from the Baiyun International Airport by car.It has 98 quiet lake scenery and deep mountain scenery well-appointed guest rooms、suites and villas,Chinese and Western restaurants,multifunctional halls,banquet halls,conference centers,fitness centers,kids clubs,outdoor pools,spa,organic farms and various types of facilities.It is an ideal venue for business meeting and leisure.

地址：广州市花都区山前大道东段北侧天湖峰境园。
Address: Sky Villa, Northeast Shanqian Da Dao, Huadu District Guangzhou, Guangdong , China.

云峰大酒店 GRANDPEAK HOTEL

地处新机场板块，距新白云机场仅15分钟车程，物流园、汽车、皮革、珠宝等产业中心驱车畅达，地理位置得天独厚。多条快速干线紧密连接广州市区及琶洲展馆。

Located in the plot of new airport, it is only 15 minutes away from the New Baiyun Airport in drive. Driving path via Logistics Park, automobile, leather, jewelry and other industrial centers is quite smooth, demonstrating the superior geographic location richly endowed by nature. Several fast main lines are closely connected with Guangzhou urban downtown and Pazhou Exhibition Hall.

地址：新华镜湖路2号　　　电话：020-86891888
Address: No. 2 Jinghu Road, Xinhua Street　　　Telephone: 020-86891888

正盛红谷酒店 ZENTH INTERNATIONAL HOTEL

位于商务金融中心区主干道迎宾路上，拥有各式豪华客房和行政套房、各种流行时尚餐厅和商务会议设施。有免费国内电话网络系统，房间配有保险箱、一体厨房、吧台等，并可享受室内高尔夫、健身及皮肤护理。

Located in the business and financial center and Yinbin Road of main stem, the Zenth International Hotel boasts luxury guest rooms and executive suites in different styles, various fashionable restaurants and business conference facilities, free domestic telephone network system, guest rooms equipped with safe box, integrated kitchen and bars, as well as indoor golf, body-building and skin treatment.

地址：迎宾大道66号正盛商务大厦　　电话：020-86902888
Address: Zhengsheng Business Building, No. 66 Yinbing Avenue
Telephone: 020-86902888

华钜君悦酒店 H. J. GRAND HOTEL

花都规模最大的商务会议型酒店，毗邻新白云国际机场、广州北站，交通十分便利，集客房、会议、餐饮、娱乐于一体，拥有设备一流的KTV及恒温泳池、健身、洗浴、桑拿，设备精良，服务细致，是商务人士的理想之选。

It is the largest business conference hotel throughout Huadu; adjacent to the New Baiyun International Airport, Guangzhou North Station, so that it possesses convenient transportation. Integrating guest rooms, conference, catering and recreation, H. J. Grand Hotel is an ideal choice for business person for KTV equipped with first-class facilities, constant-temperature swimming pool, body-building, bath and sauna, offering excellent equipment and thoughtful service.

地址：新华街迎宾大道与青莲路交界处　　电话：020-36988888
Address: Junction between Yinbin Avenue and Qinglian Road, Xinhua Street Telephone: 020-36988888

皇冠假日酒店 CROWNE PLAZA HOTEL

位于花都珠宝交易市场旁，邻近新白云国际机场；完备的康体设施、园林式户外泳池、餐厅、酒吧、多功能厅、皇冠宴会厅一应俱全，让您释然享受世外桃源。

Located close to Huadu Jewelry Trading Market and adjacent to New Baiyun International Airport. Complete recreational facilities, Crown Plaza Hotel possesses landscape-style outdoor swimming pool, restaurant, pub, multi-function room and Crowne Plaza Banquet Hall enable you to enjoy the land of idyllic beauty.

地址：新华街迎宾大道189号　　电话：020-36900888
Address: No. 189 Yinbin Avenue, Xinhua Street　　Telephone: 020-36900888

HOTEL INTRODUTION OF HUADU 花都酒店介绍

🏨 湘都大酒店 XIANGDU HOTEL

是一家集客房、桑拿、夜总会、餐厅、商务中心为一体的现代综合型酒店，毗邻广州北站，零距离繁华商业街，位置优越，是宾客商务、旅游、会议的理想会所。

It is a modern comprehensive hotel integrating guest room, sauna, night club, restaurant and business center as a whole. It is adjacent to Guangzhou North Station and prosperous commercial street, so that it is an ideal place for entertaining, business, tourist, and conference for superior location.

 地址：新华街建设北路86号　电话：020-86891288
Address: No. 86 Jianshe North Road, Xinhua Street　Telephone: 020-86891288

🏨 泛美大酒店 FANMEI HOTEL

涉外旅游商务型酒店，位于商贸繁华地段，依傍花都区政府，交通十分便利，着力为顾客营造一个温馨、舒适、超值的商务、娱乐休闲环境。

It is a foreign torist business hotel which is located in busy district and adjacent to Huadu District Government with quite convenient transportation. It strives to create a warm comfortable excellent commercial, recreational and leisure environment.

 地址：新华街公益路27号　电话：020-36888088
Address: No. 27 Gongyi Road, Xinhua Street　Telephone: 020-36888088

🏨 聚喜莱国际大酒店 JOLLIES INTERNATIONAL HOTEL

涉外型豪华酒店，将岭南文化与西式风格巧妙地揉合在一起，高雅脱俗。酒店配备餐厅、夜总会、多功能厅、水疗中心，专业贴心的管家服务使您的商务之旅回味无穷。

It is a foreign luxury hotel, skillfully combining culture of Lingnan with Western style, elegant and refined. Equipped with restaurant, night club, multi-function room, hydrotherapy center, professional and thoughtful butler service will lead to endless pleasure during your business travel.

 地址：新华街迎宾大道70号　电话：020-36968888
Address: No. 70 Yinbin Avenue, Xinhua Street　Telephone: 020-36968888